D0903674

ADAPTATION
AND
DIVERSITY

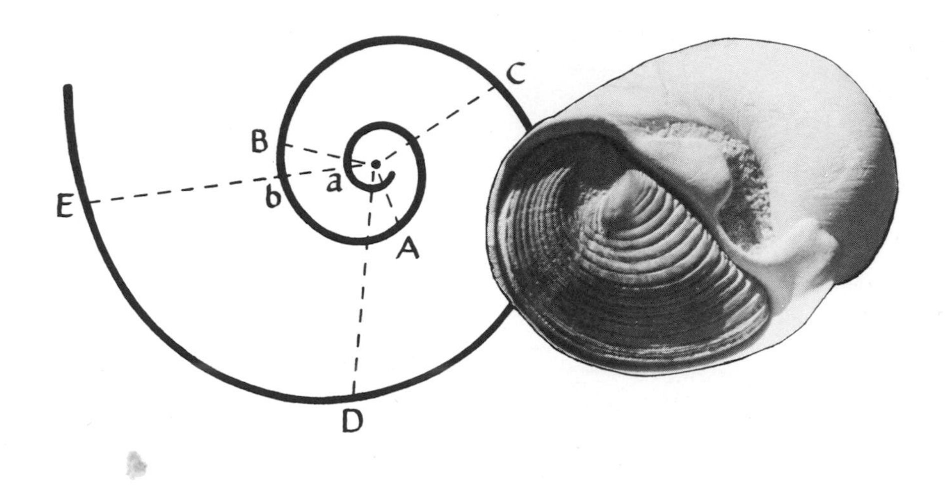

Egbert Giles Leigh, Jr.

Princeton University and
Smithsonian Tropical Research Institute

ADAPTATION AND DIVERSITY

NATURAL HISTORY AND THE MATHEMATICS OF EVOLUTION

Freeman, Cooper & Company

1736 Stockton Street, San Francisco, California 94133

Printed in the United States of America
Library of Congress Catalogue Card Number 75-124552
SBN 87735-407-3

In deepest gratitude,
this book is dedicated to four naturalists,
who have opened my eyes:

F. W. TREVOR,

G. E. HUTCHINSON,

A. G. FISCHER, and

the former E. M. HODGSON

Preface

THIS BOOK was written to justify modern evolutionary theory to the naturalist. Although it does not pretend to be a complete textbook on evolution, I trust it will prove a useful introduction to the subject. Even though the understanding of speciation is one of the triumphs of modern biology, I discuss the subject only briefly, because other topics seem more in need of emphasis at the moment. I often speak of the "perfection" of animals, not because I think most animals are perfect, but rather because this seems a useful starting point for a precise discussion of adaptation, a useful "first approximation to the truth," as it were. The discussion of population genetics is oriented toward the paleontologist, who is concerned with the long-term prospects of different genetic "strategies," and who needs to know in what sense, if any, selection is oriented toward "the good of the species"; most population geneticists would summarize the subject very differently.

The bibliographic notes point out some of the biases of the text. I think the book might well be used in conjunction with Ehrlich and Holm's *The Process of Evolution*, or G. Ledyard Stebbins's *Processes of Organic Evolution*.

The unfailing kindness of many people has made this book possible. I began it at Stanford, where Ehrlich invited me to visit the spring and summer of 1966: I remember with pleasure writing day after still summer day in an arbor outside the Stanford Museum. Several winters at Princeton, devising lectures for a course in evolution, gave the book its present form; the department very kindly relaxed the demand for professional papers during this period.

The past three years have also introduced me to the tropics. I have spent several Christmases on Barro Colorado Island, in the midst of the Panama Canal, where one could watch hummingbirds at play in the garden or the weather building up in the hills across the lake. The tropics are a strange experience: it takes some while to sense the profound order in the seeming riot of vegetation, and to realize how much more prevalent this type of vegetation is than Pennsylvania's. Our perspective needs correcting when we gawk at palms and take the much more bizarre pines for granted.

The Smithsonian, the American Philosophical Society, and Princeton University jointly staked my wife and myself to a half-year's trip around the world, which cannot help but open one's eyes: Madagascar, Australia, and New Guinea are strange worlds which set the familiar in perspective. Thanks to M. Peyrieras, I wrote some of this book in one of the last tracts of rain forest remaining in Madagascar, a forest where one could talk to lemurs every evening. Now the Smithsonian Tropical Research Institute enables me to spend half my time in the tropics, so the book which began in California was completed in a tropical forest.

Mr. Stephen Rawson and Dr. Donald Baird kindly provided some splendid photographs. This book also owes much to conversations with my Princeton colleagues, especially A. G. Fischer, Henry Horn, and Robert MacArthur, who have also read drafts of many of the chapters.

EGBERT G. LEIGH, JR.

Barro Colorado Island, Canal Zone
January, 1971

Contents

List of Illustrations

ADAPTATION AND DIVERSITY

CHAPTER I

Introduction

THIS BOOK is intended as a brief description of evolutionary theory or, perhaps more accurately, as a description of ideas which might be useful in formulating such a theory.

We must make clear at the outset what the theory is to explain, and how it can be verified; else we shall risk falling into the most appalling confusion. The cardinal problem of evolution is to explain the origin and nature of adaptation; we begin, therefore, by showing how to recognize the perfection of living things each for its way of life. Since organisms are adapted to specific roles, the study of adaptation quickly forces us to enquire into the relationship between different ways of life: we cannot truly understand adaptation unless we understand it in the context of the balance of nature. The best evidence concerning evolution as an *adaptive* process lies in the fossil record, so we shall enquire into the character, and degree of detail, of the story it has to tell, thereby setting our study of evolutionary mechanisms in proper perspective.

The stage is finally set, and we can turn to the workings of the evolutionary play. We shall enquire into the mechanism of natural selection and discuss the sorts of adaptation it can lead to. This requires some genetics, since the evolution of a population involves some change in its hereditary make-up.

I have tried as far as possible to draw my examples from the world of familiar observation, rather than from elaborate experiment. It is remarkable how much one can learn by rightly looking at field and forest, seashore and mountainside. It is an easy business to guess that little desert shrubs, which cast hardly any shade at all, are evenly spaced by competition for water rather than for light. Having made the guess, one finds it an intriguing matter to decide what else to observe to check this explanation. Again, it is a frequent exercise in ecology courses to try and piece together the stages of a plant succession by asking what seedlings grow under what adult plants. One can then ask what interrupts normal plant succession to maintain a pine barren or, turning the question about, one can ask how the pine barren community maintains itself. A little knowledge of soil properties and of the effects of fire permits one to frame some quite reasonable guesses.

Mathematics extends our power to read nature in this fashion quite a long way. They are quite simple and general principles which tell us what is to be learned about environmental conditions from the zonation of different-sized plants on a mountainside, or the distribution of deciduous and evergreen trees in a ravine. Thus I hope this book will make the book of nature, as well as the theory explaining its origins, a little easier to understand. Sadly, the natural world is easiest to interpret when it has had time to adjust fully to its environment: now that man is proving an ever harsher ruler of his world, such conditions are rarely met. Still, I hope this book is not entirely a reminiscence of things past.

I will rely heavily on mathematics as a means for both

testing and developing hypotheses. Sometimes the idea we wish to develop is a mathematical one, as when we describe the form of sponge or snail shell. Here, numerical calculations are needed to verify the hypothesis. Mathematics here plays the role freshman physics taught us to expect: many a reader has tested Boyle's Law by comparing a series of numbers obtained in laboratory experiment with a series predicted from an equation. In the later parts of the book, we will also use mathematics as an aid in developing hypotheses, just to make sure our arguments lead to the conclusions we think they do. Ecology and evolution involve many complicated topics, easily obscured by the misuse of words. Mathematics can also confuse, but once one becomes familiar with it, mathematics can serve as a torch in the darkness, illuminating the problems somewhat, and even revealing answers on occasion.

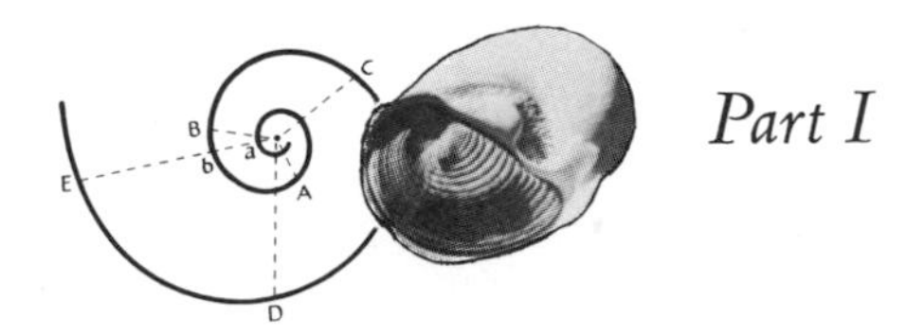

Part I

ADAPTATION

Living creatures are so obviously adapted *that previous generations felt a Creator must have perfected each and every one of them for its station in life. Evolution is the study of the origins of adaptation; before embarking on this, we must learn how to recognize and measure adaptedness. In doing so we find we can learn rather a lot about the process of evolution by studying its "finished products."*

CHAPTER 2

The Perfection of Sponges

A SIMPLE sponge is a chamber with a major opening, the *osculum*, at the top and pores in the sides. The interior of the chamber is lined with flagellated *collar cells*, which continually beat the water with their flagella, as if trying to swim into the chamber's wall. These "swimming motions" draw water through the pores into the chamber, from whence it is forced out through the osculum. The resulting current carries food particles into the sponge, where they are trapped and eaten by the collar cells, and bears away wastes of metabolism and respiration through the osculum. In still water, the "suction" of the flagella will create currents drawing ejected water back to the sponge. The perfect sponge must expel filtered water as far away as possible, so that the return currents can dissipate wastes and acquire new food particles. How should the sponge do this?

The surrounding water slows the velocity of a sponge's jet, just as friction slows a block sliding across a table. If several sponges are united so that their excurrent jets lie side by side, the area of contact between these jets and the surrounding water is reduced: united, the jets carry farther than any would alone. Such compound sponges (Fig. 2-1) are quite common. To calculate their advantage, notice that the velocity of a sliding block is reduced continually, by a fixed proportion each second: if the block moves v inches in one second, it moves $v(1-u)$ in the next, $v(1-u)^2$ in the third, and so on. The total distance travelled is $v[1+(1-u)+(1-u)^2+\ldots]$, or v/u inches: the distance travelled is proportional to the initial velocity v, and inversely proportional to the resistance u the table offers the block*. Experiment tells us that if the sponge's jet is narrow

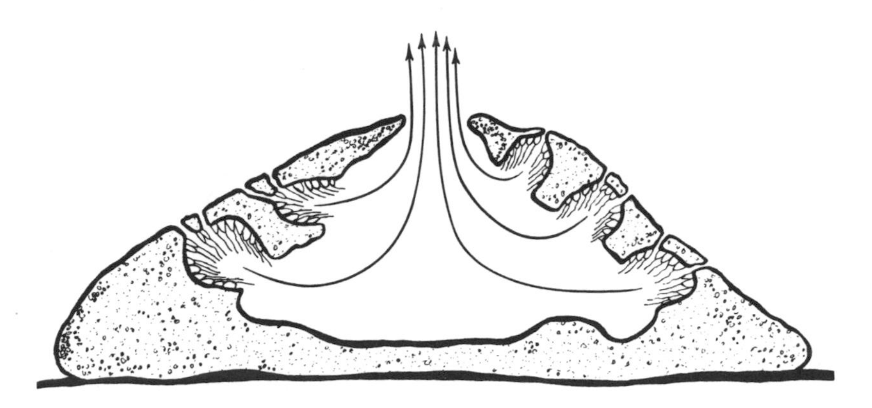

Fig. 2-1. A rhagon, made up of several "simple sponges" so arranged that their excurrent jets are united. (After text-figure 5, p. 298, of G. P. Bidder, "The Relation of the Form of a Sponge to its Currents," *Quarterly Journal of the Microscopical Society*, vol. 67, 1923, with permission of the Microscopical Society.)

---

* Notice that $1+x+x^2+x^3+\ldots$ multiplied by $1-x$ is 1, so $1+x+x^2+x^3+\ldots$ must be $1/1-x$. All this assumes, of course, that x is smaller than 1.

enough to be coherent, it, too, will carry a distance proportional to its initial velocity, and inversely proportional to the resistance from its surroundings. If we imagine our jet sliced, like a loaf of French bread, we would find the resistance on each segment proportional to the ratio of the area of its exposed surface to its weight. The area of contact between each segment and the surrounding still water is proportional to the jet's diameter, while the segment's weight is proportional to the square of the jet's diameter: the length of the jet is thus proportional to its width.

How can our compound sponge form the longest possible jet? Consider the analogy with a garden hose. The collar cells maintain a pressure within the sponge analogous to the hydrostatic pressure behind the spigot to which the hose is attached: in each case, the pressure moves the water. If the hose is fully open, the water falls limply upon the ground; it moves so quickly through the hose that it dissipates as friction against the sides of the hose much of the potential energy implicit in the hydrostatic pressure behind the spigot, and the kinetic energy of the emerging water is low. Partly close the hose-opening with a thumb, and the water carries much farther: it cannot move so quickly through the hose, so less energy is lost as friction, and its kinetic energy is greater. Since less water emerges, the water's velocity must be so much the greater. Close the opening nearly all the way, however, and the stream, for all its energy, will be too thin and light to travel far. An intermediate opening, balancing the advantages of a weighty stream against those of a speedy one, provides the jet which carries farthest. So with the sponge, where an intermediate-sized osculum forms the longest jet.

To learn what size osculum ejects water farthest, we must calculate the precise effect of oscular diameter on jet length. Jet length is proportional to the diameter of the osculum and to the speed of the emerging water; the speed is determined in turn by the energy the flagella impart to this water. The kinetic energy of each second's flow of water is the force acting on it at the osculum, times the velocity at which it is ejected. Dividing the force and multiplying the velocity by the cross-sectional area of the osculum, we find that this energy is the pressure acting at the osculum, multiplied by the volume q of a second's flow of water (remember that pressure is force per unit area). Were there no friction, this would be Pq, where P is the pressure the collar cells maintain in the flagellar chambers, and q is the volume of water ejected per second.

We use the theory of streamlined flow to calculate the energy lost as friction against the interior canals and chambers of the sponge. Perhaps those of you who have watched smoke rising from a lighted cigarette have noticed how it sometimes "flows" in streamlines before it curls into vortices and diffuses through the room. Liquid flow can be equally streamlined. Imagine that liquid flows through the pipe as a nested series of concentric shells, like the cylinders of a folding telescope. If a steady pressure forces liquid through the pipe, the outermost shell of fluid, next to the pipe's still surface, moves very slowly, while the inner ones move faster: the velocity of each shell of fluid balances the pressure which moves it against friction with the shell's slower outside and faster inside neighbors. The friction between neighboring cylinders represents the viscosity, or "stickiness," of the fluid. Assuming the canals and chambers of the sponge are so many little pipes, we find (see Appendix 1) that the energy lost in them is Fq^2, where

$$F = 8\pi n[N_1 b_1/a_1{}^2 + N_2 b_2/a_2{}^2 + N_3 b_3/a_3{}^2 + b_4/a_4{}^2],$$

$\pi = 3.1416$, and n is the viscosity of water. N_1, N_2, and N_3 are the numbers of afferent canals, flagellar chambers, and efferent canals; b_1, b_2, and b_3 their average lengths; and a_1, a_2, and a_3 their aggregate cross-sectional areas. b_4 is the average distance a water molecule travels in the cloaca and a_4 is the cloaca's cross-sectional area.

To find the velocity of the water ejected by an osculum of radius R, we use the equation for the kinetic energy of a second's flow of water,

1) $Pq - Fq^2 = (1/2)\rho q v^2.$

ρ is the density of water, q the volume ejected per second, v the water's velocity, P the flagellar pressure, and F the "coefficient of friction." Setting q equal to $\pi R^2 v$, where πR^2 is the cross-sectional area of the osculum, we obtain

2) $P - F\pi R^2 v = \rho v^2/2.$

To calculate the length L of this sponge's jet, use the equation expressing the proportionality of L to the jet's velocity and the diameter of the osculum,

$$L = CRv.$$

C is a constant of proportionality which must be measured experimentally. Substituting $v = L/CR$ into equation 2, we obtain

$$P - F\pi RL/C = (1/2)\rho L^2/C^2R^2,$$

or, more simply,

3) $PR^2 - F\pi R^3 L/C = (1/2)\rho L^2/C^2.$

Solving for L tells how long a jet is formed by an osculum of radius R.

To calculate the best oscular diameter, we take the differential of equation 3, obtaining

$$2PRdR - (3\pi F/C)LR^2dR - (F\pi/C)R^3dL = \rho LdL/C^2.$$

We seek that value of R for which $dL = 0$, for which L is maximum (see Appendix 2). $dL = 0$ implies

$$2PR - (3\pi F/C)LR^2 = 0;$$

$$L/C = 2P/3\pi FR.$$

Substituting for L/C in equation 3, we obtain

$$PR^2 - 2PR^2/3 = 2P^2\rho/9\pi^2F^2R^2;$$

4) $$R^4 = 2P\rho/3\pi^2F^2.$$

How do we know whether a sponge's osculum is adjusted to form a long jet? Bidder, who first analyzed the perfection of sponges, studied one in detail to see how large its osculum should be. He counted and measured its canals to calculate F, and measured the volume and speed of the water it ejected to find P from equation 1; knowing this, he could calculate the optimum R from equation 4 (see Problem 1). The sponge's osculum seemed close to the right size, but it was hard to be sure, for there was so much room for error in the measurements.

Bidder also devised more reliable tests. If flagellar pressure weakens, the oscular diameter should shrink in proportion to the square root of the velocity of the sponge's diminishing jet (Problem 2). And, indeed, a muscular sphincter contracts the osculum in just this manner when water pressure slackens. Under normal circumstances, however, P will be the same for all sponges of the same species, for it depends only on the nature of the collar cells and the shape of their chambers. Thus, among sponges of the same kind, the cross-sectional area of the osculum should vary in proportion to the sponge's volume: doubling the sponge's size halves F, for the number of canals and their aggregate cross-sectional area are both doubled, and, by equation 4, doubles R^2. Moreover, sponges of the same kind should eject water equally fast, as halving F and doubling R^2 leaves the equation for jet velocity (equation 2) unaltered. Bidder found all this true for most of the sponges he studied. A large bath sponge, however, has several oscula, each a centimeter in diameter, because a single larger osculum would form an incoherent jet.

One cannot speak of perfection in the abstract. What is

perfect for one purpose is useless for another: perfection must be discussed with respect to a goal. The sponges living in turbulent places, or in the ocean depths where the waters' subtle creep brings food and bears wastes away, are open, sometimes fan-like (Fig. 2-2), to permit the free passage of the current; they are quite different from sponges perfected to still water. Thus we may deduce something of a sponge's habitat from its form. We will learn later how one might interpret the habits and habitats of fossils by reading such riddles of form.

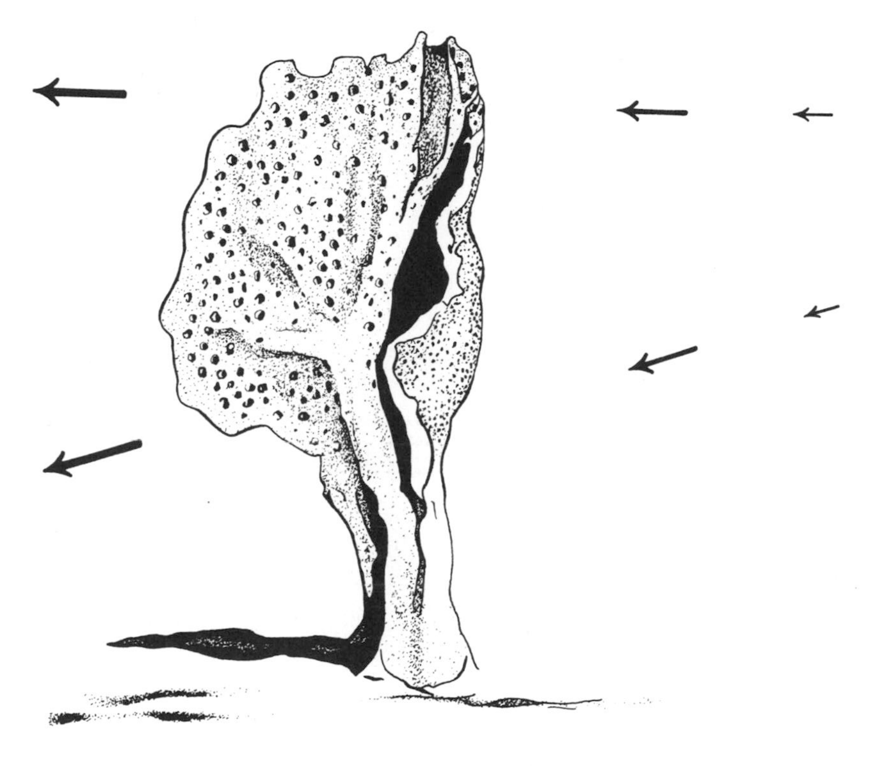

Fig. 2-2. A fan sponge, *Phakellia granulosa*, with arrows showing direction of current. (After text-figure 12 of Bidder, *op. cit.*, p. 311, with permission of the Microscopical Society.)

Appendix 1 – Losing Energy in a Pipe

Consider a fluid moving in streamlines through a pipe of length L and radius R. Imagine the fluid partitioned into concentric cylindrical shells of thickness Δr, each moving faster than its outside neighbor. A longitudinal cross-section of the pipe and its contents is diagrammed in Fig. 2-3; shell A, in contact with the still walls of the pipe, moves the slowest, B faster, and C faster still. The velocity of each shell of fluid is set by the balance between the pressure forcing the liquid through the pipe and the frictional force exerted by neighboring shells.

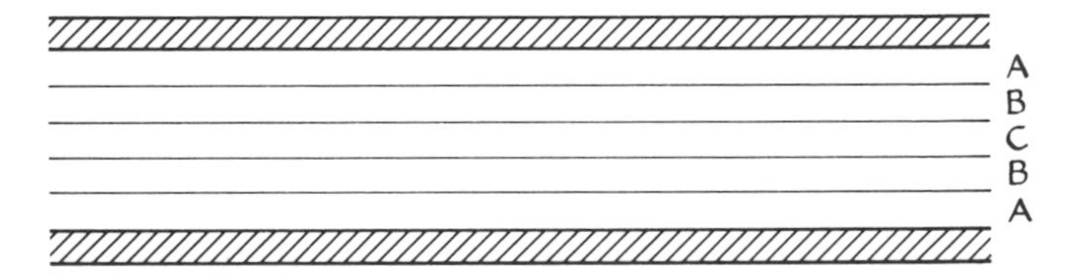

Fig. 2-3. Cross-section showing laminar flow of fluid through a pipe. The fluid flows as concentric "shells": shell A, in contact with the walls of the pipe, flows most slowly, while the middle one flows fastest.

Consider now a cylindrical shell of inner radius $r - \Delta r$ and outer radius r; let its velocity be $v(r)$, that of its outer neighbor be $v(r + \Delta r)$, and that of its inner neighbor be $v(r - \Delta r)$. The frictional force exerted on this shell by its outside neighbor is proportional to the area of the zone of contact between them, and to the ratio of the difference $v(r + \Delta r) - v(r)$ in their velocities to the thickness Δr of the shells. The frictional force in question is thus

$$-2\pi n r L \left[\frac{v(r + \Delta r) - v(r)}{\Delta r}\right].$$

The negative sign indicates that this force acts in the direction opposite to velocity. The "constant of proportionality" n measures the fluid's viscosity. The frictional force exerted by our shell's inside neighbor is

$$2\pi n(r - \Delta r)L\left[\frac{v(r) - v(r - \Delta r)}{\Delta r}\right].$$

If there is a constant pressure forcing the fluid through the pipe, it exerts a force

$$P\pi[r^2 - (r - \Delta r)^2],$$

or roughly $2\pi Pr\Delta r$, on our shell. The balance between pressure and friction is expressed by the equation

5)
$$2\pi Pr\Delta r + 2\pi nL\left\{r\left[\frac{v(r + \Delta r) - v(r)}{\Delta r}\right] - (r - \Delta r)\left[\frac{v(r) - v(r - \Delta r)}{\Delta r}\right]\right\} = 0.$$

Mentally partitioning the fluid into shells is inaccurate in one respect: although all the fluid the same distance from the center of the pipe moves at the same speed, this is not true for the fluid in a shell of finite thickness; fluid velocity varies continuously with distance from the center of the pipe. Thus we must divide equation 1 by Δr and take the limit as Δr passes to zero. Since, for any function f,

$$\lim_{\Delta r \to 0} \frac{f(r) - f(r - \Delta r)}{\Delta r} = \frac{df}{dr},$$

we may set

$$\lim_{\Delta r \to 0} \frac{1}{\Delta r}\left[r\frac{v(r + \Delta r) - v(r)}{\Delta r} - (r - \Delta r)\frac{v(r) - v(r - \Delta r)}{\Delta r}\right]$$

equal to

$$\frac{d}{dr}\left(r\frac{dv}{dr}\right).$$

Substituting into equation 1, we obtain

$$2\pi Pr + 2\pi nL\frac{d}{dr}[r(dv/dr)] = 0;$$

$$Pr/nL = -\frac{d}{dr}[r(dv/dr).$$

Integrating, and calling A_1 the resulting constant of integration, we obtain

$$Pr^2/2nL - A_1 = -r(dv/dr).$$

Solving for dv/dr, we obtain

$$dv/dr = -Pr/2nL + A_1/r.$$

Integrating again, and calling A_2 the new constant of integration, we obtain

$$v(r) = -Pr^2/4nL + A_1 \log r + A_2.$$

The velocity of the fluid at the center of the pipe is obviously finite: since the logarithm of 0 is infinite, A_1 must be 0. Since the fluid touching the walls of the pipe is still,

$$v(R) = -PR^2/4nL + A_2 = 0;$$

$$A_2 = PR^2/4nL.$$

We thereby obtain

$$v(r) = P(R^2 - r^2)/4nL.$$

To find the volume of a second's flow of liquid through the pipe, we multiply the velocity $v(r)$ of each shell of liquid in the pipe by the cross-sectional area $2\pi rdr$ of the shell, and add all these terms together. We may represent this sum by the integral

$$\int_0^R 2\pi rv(r)dr = (\pi P/2nL)\int_0^R (R^2r - r^3)dr = P\pi R^4/8nL.$$

The work done per second to keep the fluid moving at this speed, and hence the rate at which the fluid's energy is dissipated, is the pressure P acting on the fluid times the volume q of fluid moved per second, or Pq. Since q is $P\pi R^4/8nL$, the pressure P required to move q units of water through the tube each second is $8qnL/R^4\pi$. Setting the cross-sectional area πR^2 of the pipe equal to a, we obtain

$$P = 8\pi nLq/a^2,$$

$$Pq = 8\pi nLq^2/a^2.$$

The energy lost per second by fluid flowing in parallel through m such pipes is m times that lost in a single pipe, or

$$m[8\pi nLq^2/a^2] = m[8\pi nL(mq)^2/(ma)^2].$$

Setting mq, the total volume of fluid emerging from these pipes per second, equal to Q, and letting ma, the aggregate cross-sectional area of these pipes, equal A, the energy loss may be expressed as $8\pi nmLQ^2/A^2$, as stated in the text.

Appendix 2 – The Use of Differentials

In taking the differential of both sides of equation 3, we are asking what effect a small change in the diameter of the osculum will have on the velocity of the jet. If oscular radius and jet length are R and L, respectively, what change ΔL in the jet length will result from a change ΔR in the oscular radius? R and L are related by the equation

6) $$PR^2 - F\pi R^3 L/C = (1/2)\rho L^2/C^2.$$

$R + \Delta R$ and $L + \Delta L$ must be related by the same equation: thus

7) $$P(R + \Delta R)^2 - F\pi(R + \Delta R)^3\,(L + \Delta L)/C = (1/2)\rho(L + \Delta L)^2/C^2.$$

If we assume ΔL and ΔR are so small that $(\Delta L)^2$, $(\Delta L)(\Delta R)$, and $(\Delta R)^2$ are all effectively zero, our second equation becomes

$$8) \quad P(R^2 + 2R\Delta R) - F\pi(R^3 L + 3R^2 L\Delta R + R^3 \Delta L)/C$$
$$= (1/2)\rho(L^2 + 2L\Delta L)/C^2.$$

If we subtract the first equation from the third, we obtain an equation relating ΔR and ΔL, to wit:

$$9) \quad 2PR\Delta R - F\pi(3R^2 L\Delta R + R^3 \Delta L)/C = \rho L\Delta L/C^2.$$

Imagine now that $R = 0$, and consider what happens when we increase it by successive small steps of magnitude ΔR. At first, increasing R increases jet length, but as we approach the optimum radius these increases ΔR will produce successively smaller increases in jet length. Once R passes the optimum, further increases will produce successively greater *decreases* in the jet's length. The optimum oscular radius is that for which an increase ΔR no longer increases jet length but has not yet begun to decrease it. Mathematically, the optimum R is that for which $\Delta L/\Delta R = 0$, that is to say, for which $\Delta L = 0$.

If we set $\Delta L = 0$ in equation 9, we obtain

$$10) \quad 2PR\Delta R - 3\pi FR^2 L\Delta R/C = 0,$$
$$2PR = 3\pi FR^2 L/C,$$

at which point we have returned to the world of algebra.

Problems

1. Consider a sponge with 81,000 afferent and 5,200 efferent canals. The average lengths of an afferent and an efferent canal are .06 cm. and .08 cm., respectively; the total cross-sectional area of afferent and efferent canals, respectively, is 4.2 and 2.5 cm^2. The cloaca is 10 cm. long, and its cross-

sectional area is .21 cm^2. When the osculum is .28 cm. wide, its jet emerges at 8.5 cm./sec. (data from Bidder). What is the optimum diameter of this osculum? In your answer, ignore friction against the flagellar chambers.

Solution: To find the optimum diameter of the osculum, we must know F, the "friction coefficient," and P, the pressure maintained in the flagellar chambers.

$$F = 8\pi n[N_1 b_1/a_1{}^2 + N_2 b_2/a_2{}^2 + N_3 b_3/a_3{}^2 + b_4/a_4{}^2].$$

$8\pi n = .28$; N_1, N_2, and N_3 are the numbers of afferent canals, flagellar chambers, and efferent canals; b_1, b_2, and b_3 their average lengths; and a_1, a_2, and a_3 their total cross-sectional areas. b_4 is the average distance a water particle travels in the cloaca, which we assume to be half its length, and a_4 is the cloaca's cross-sectional area. We assume $N_2 b_2/a_2{}^2 = 0$, which affects our answer less than one per cent. Substituting the required numbers,

$$F = .28[81{,}000 \times .06/(4.2)^2 + 5{,}200 \times .08/(2.5)^2 + 5/(.21)^2],$$

or 128.

To calculate the pressure P maintained in the flagellar chambers, we use the equation

$$P = \rho v^2/2 + F\pi x^2 v,$$

where x is the radius of the osculum when water is ejected at v cm./sec., and ρ is the density of sea water, which is nearly 1. v is 8.5 cm./sec., so $\rho v^2/2$ is 36. $x = .14$ cm., so $\pi x^2 = .062$ $cm.^2$, and $\pi F x^2 v = 128(.062)(8.5) = 67$. Thus $P = 103$.

The optimum oscular radius R is given by the equation

$$R = \sqrt[4]{2P\rho/3\pi^2 F^2} = \sqrt[4]{206/490{,}000}.$$

This is .143 cm. Since F can only be calculated to an accuracy of about 15% (it is rather hard to be sure of the aggregate cross-sectional area of 81,000 afferent canals), the optimum value does not differ significantly from the observed.

2. Show that if the pressure in the flagellar chambers drops, the diameter of the osculum should decline as the square root of the velocity of the jet.

Solution: The jet's velocity can be found from the equation

$$P - F\pi R^2 v = \rho v^2/2,$$

where P is the pressure in the flagellar chambers, R is oscular diameter, v is jet velocity, and ρ and F are constants. If R is optimum,

$$R^4 = 2P\rho/3\pi^2 F^2;$$
$$R^2 = (1/\pi F)\sqrt{2P\rho/3}.$$

Substituting for R^2 in our first equation, we obtain

$$P - v\sqrt{2P\rho/3} = \rho v^2/2,$$

$$v = \frac{-\sqrt{2P\rho/3} + \sqrt{8P\rho/3}}{\rho} = \sqrt{2P/3\rho}.$$

Since the optimum oscular diameter varies as the fourth root of pressure, it varies as the square root of velocity.

Bibliographical Notes

This chapter is based on G. P. Bidder's "The Relation of the Form of a Sponge to its Currents," pp. 293-325 of the *Quarterly Journal of the Microscopical Society*, vol. 67, 1923. This paper is a pleasure to read, and deserves to be far more widely known. Bidder explores the subject further in "The Perfection of Sponges," pp. 119-146 in the *Proceedings of the Linnean Society,* vol. 149, 1937. A discussion of laminar flow in a tube occurs under the heading of Poiseuille's Law in C. Tanford, *Physical Chemistry of Macromolecules*, Wiley, 1961.

To speak of the perfection of animals is to be accused of believing that this is "the best of all possible worlds." How can it be, when it is still evolving? For contrasting views on this subject, see C. S. Pittendrigh, "Adaptation, Natural Selection and Behavior," pp. 390-416 in A. Roe and G. G. Simpson, eds., *Behavior and Evolution*, Yale University Press, 1958, and A. J. Cain, "The Perfection of Animals," pp. 36-63 in J. D. Carthy and C. Duddington, *Viewpoints in Biology,* vol. 3, 1964.

Indeed, the question of perfection imparts a strange ambivalence to evolutionary theory (see, for example, the essays of Bohm and Grene in C. H. Waddington, ed., *Towards a Theoretical Biology*, vol. 2, Aldine, 1969). It is the oldest problem of biology; it preoccupied Aristotle and the Schoolmen. The perfection of living things was long thought to prove the existence of God: what mechanical process could provide purpose in life for all creatures? Imagine the comedown when Darwin suggested that living things were organized only to reproduce, and that their organization was the product of the automatic mechanism of natural selection!

Physiologists remind us that it is not enough to determine the roles for which organisms are perfected, and to predict the adaptations which these organisms possess. We cannot put our ideas of perfection on an honest footing until we trace, step by step, how this organization arose, analyze its function, and thereby learn in detail how it can be adjusted by natural selection. Analysis of this sort often uncovers an unmanageable welter of detail. By contrast, emphasis on perfection yields greater perspective, but it becomes all too tempting to ascribe a character to the omnipotence of natural selection (as older generations ascribed it to God) without understanding what the character is good for.

CHAPTER 3

Why Are There as Many Men as Women?

IN THE last chapter we discussed the form and function of a sponge, demonstrating its perfection for its way of life. Our argument fell into two parts: an ecological part showing that the sponge must eject filtered water as far away as possible and a physiological part showing how the sponge accomplished this. The ecology was fairly easy; the physiology, however, was rather involved and mathematical. Indeed, we could only "solve" the physiology by flaking off an easy aspect of the problem which could be mathematized: we related the length of a jet to the diameter of an osculum, suppressing the remaining structural aspects of the problem into a pair of constants. We thus determined the optimum diameter of the osculum without addressing ourselves to the perfection of collar cells, flagellar chambers, and the like, which in our present state of ignorance we are in no wise capable of discussing.

In general, then, there are two problems in establishing the perfection of an organism. There is the question of the criteria of perfection: how do we compare the fitness of different individuals? The Bible teaches us to judge a tree by its fruit; correspondingly, we judge the perfection of an organism by its power to survive and multiply. The criteria of perfection thus come from ecology and genetics. In principle, at least, these criteria are easy to find, and in this book we shall suggest guidelines for the search. Comparison, however, presupposes a series of alternatives to be compared. It does not do merely to have a criterion by which we may judge the fitness of an organism, or of a species; we must imagine the array of alternative forms, structures, or strategies the species might have had. To describe these alternatives definitively enough to be able to compare their merits is a very difficult, often seemingly impossible, problem. It is so much easier to pick out the best from an array of possibilities than to find out what the possibilities are!

The World of Graphs and Contour Lines

There is a formal procedure for establishing perfection. Take as sample problem the question of why there are as many males as females in most populations. Does an individual maximize his contribution to the genetics of future generations by raising as many sons as daughters?

It is easy enough to delimit the array of possible families. Represent a family of x sons and y daughters as a point (x, y) on a plane graph (Fig. 3-1). An individual can bring up only a limited number of children: shade the portion of the graph representing families he could afford to raise. If it is impossible to bring up more than k children, and if sons and daughters are equally easy to raise, the graph will be shaded between the origin and the line $x + y = k$. To compare the value of different families, we plot a series of contour lines on the graph. Each contour is a locus of points representing families which will contribute equally to the genetics of future generations. The highest contour line touching the shaded area represents the

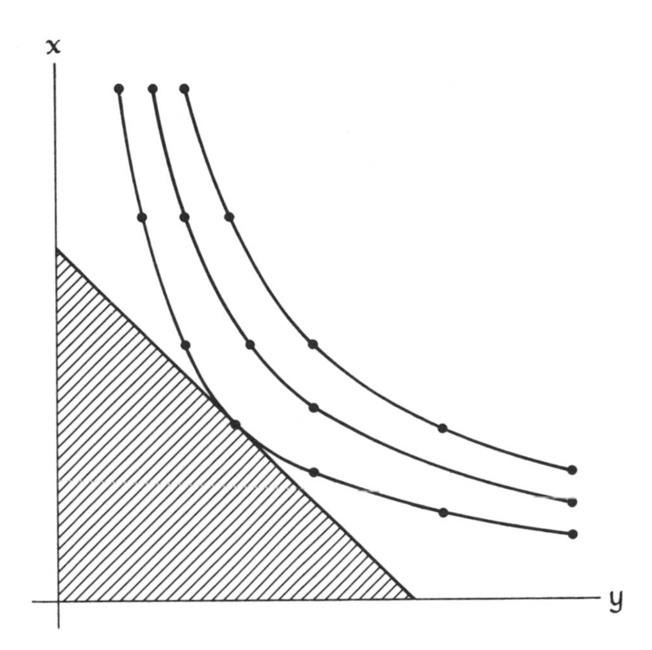

Fig. 3-1. How to calculate the optimum family composition. The straight line bounding the shaded part of the graph is the "budget line," which delimits families one can afford to raise. The hyperbolae are "indifference curves": each curve represents families of equal value to their progenitors. The point where the innermost hyperbola touches the budget line represents the best family.

greatest genetical contribution it is possible to make; the point where this contour touches the shaded area represents the family which will make it.

How do we derive our contours of equal value? Think of the genetic contribution of a family as a return on an investment of effort. Because every individual is the offspring of a male and a female, the two sexes contribute equally to future generations: the total investment of the population in each sex gives the same return. If the population as a whole devotes more effort to raising females, an individual can secure greater return on *his* investment by raising males, for his contribution will be less diluted by those of others. Thus natural selection balances the effort devoted to raising each sex: in the perfect population each individual devotes equal effort to raising males and females.

Mathematically, this means that the product xy of the number of sons and the number of daughters is maximized: the greater xy, the greater the family's contribution. To see this, observe that if A sons can be raised in place of one daughter, the limit on possible families takes the form $x + Ay = k$. xy, or $y(k - Ay)$, is maximized when $y = k/2A$, $x = k/2$. If daughters are A times as difficult to raise as sons, then one should raise A times as many sons as daughters, which is just another way of saying equal *effort* should be devoted to each. Contours representing families of equal value are thus given by the curves $xy =$ const. (other proofs of this are given in the Appendix).

Such graphs are a convenient pictorial means of representing possible alternatives and their relative worth. Biologists like pictures, and this method accordingly finds frequent use. Any time organisms can be described in terms of two essential features, such as fitness in each of two environments, or effectiveness in using each of two resources, the possibilities may be represented as the shaded portion of a graph whose two coordinates are the features in question: the best possibility is that touching the contour of highest value. These graphs, moreover, are not unique to biology. You who have taken economics have seen graphs on which are drawn indifference curves, contours of purchases yielding equal satisfaction to the buyer, and a budget line bounding the set of purchases the buyer can afford.

The World of Quarrelsome Males

Can one claim that there are as many men as women because meiosis in sperms guarantees that half the spermatozoa carry male chromosomes and half female? Boys have a higher natural death rate than girls. Thus a male birth costs a family less, on the average, than a female one, because the boy is more likely to die early and save his family the cost of raising him to maturity. On the other hand, these deaths imply that the cost for each boy raised is greater than that for girls. If equal effort is to be devoted to raising boys and girls, more boys should be born than girls, but the excess should be small enough that more girls are raised than boys. Among humans, 105 boys are born for every 100 girls. Death rates have so changed recently that this "neonate sex ratio" no longer ensures that more girls are raised than boys; still, the imbalance in the neonate sex ratio suggests that it is not predetermined by the symmetry of meiosis, but is adjusted by natural selection.

The nearly equal numbers of the two sexes show that the individual's contribution to the genetics of his species determines the survival of his type in the population, whether or not this benefits the population as a whole. A population might very well grow faster if it consisted mostly of females: after all, they bear and raise the children, while all too often the males lag superfluous upon the stage, fighting each other or amusing themselves in other ways that do not serve the population's good. For this reason, the farmer maintains many cows but only a few prize bulls. In some species, the male dies shortly after mating. Certain spiders show the most acute understanding of the case; the female eats the male shortly after mating time. But is is the well-nigh universal rule that there be as many males as females in a population: most departures from this rule are due to differences in death rates between the two sexes. Quarrelsome males are the price the population pays for sexuality, and such are the advantages of sexual reproduction that the great majority of species readily pay this seemingly frightful tax for it.

Appendix – Drawing Contour Lines

In the text, we derived contours of equal value from the assumption that equal effort should be devoted to raising males and females. We shall now present the evidence for this assumption.

Consider an individual of generation t who has x sons and y daughters. What contribution does he make to generation $t + 2$? Generation $t + 1$ will include his x sons, X men not his sons, his y daughters, and Y women not his daughters. The members of generation $t + 2$ derive half their genes from the men, and half from the women, of generation $t + 1$. The genetic contribution q of our t^{th} generation individual is therefore measured by the average of the proportion of his sons among the men of generation $t + 1$ and the proportion of his daughters among the women thereof. In symbols,

$$q = (1/2)\left[\frac{x}{x + X}\right] + (1/2)\left[\frac{y}{y + Y}\right].$$

We have a choice of means for deriving contours of equal

value: a quick method involving some sleight of hand which may invalidate it, and a correct but rather involved procedure.

First for the quick and dirty: a contour of equal value is described by the differential equation $dq = 0$, or

$$d\left[\frac{x}{x+X}\right] + d\left[\frac{y}{y+Y}\right] = 0.$$

As we are only concerned with varying x and y, this equation yields

$$\frac{dx}{x+X} - \frac{xdx}{(x+X)^2} + \frac{dy}{y+Y} - \frac{ydy}{(y+Y)^2} = 0,$$

$$Xdx/(x+X)^2 + Ydy/(y+Y)^2 = 0,$$

$$-\frac{dx}{dy} = \frac{Y}{X}\left[\frac{X+x}{Y+y}\right]^2.$$

If the perfect strategy has spread through the entire population, as indeed it should, then $x/y = X/Y = (x+X)/(y+Y)$. The equation $dq = 0$ becomes

$$dx/dy = -y/x,$$

or

$$ydx + xdy = d(xy) = 0.$$

If the family's strategy is shared by the entire population, contours of equal value are given by the curves xy = const. The sleight of hand lies in when we set $x/y = X/Y$: if we had done so before taking the differential of q, we would have found ourselves differentiating a constant. One is somehow left with the feeling that the proof depends on choosing the right time to divide by zero.

The second approach asks what effect natural selection has on family composition. Suppose the limit on possible families is given by $x + Ay = k$, and suppose that in the population as a whole $A + z$ sons are raised for each daughter, so that $X = (A + z)Y$. Will a family of "optimum" composition make a greater contribution than one of average composition? In other words, will a family with $k/2$ sons and $k/2A$ daughters make a greater contribution than one with $k(A + z)/(2A + z)$ sons and $k/(2A + z)$ daughters? The contribution q of an optimum family will be

$$q = (1/2)\left[\frac{k/2}{k/2 + (A + z)Y}\right] + (1/2)\left[\frac{k/2A}{k/2A + Y}\right]$$

$$= (1/2)\left[\frac{1}{1 + 2Y(A + z)/k} + \frac{1}{1 + 2AY/k}\right].$$

The contribution q' of an average family will be

$$q' = (1/2)\left[\frac{k(A + z)/(2A + z)}{\frac{k(A + z)}{2A + z} + (A + z)Y} + \frac{k/(2A + z)}{Y + k/(2A + z)}\right]$$

$$= (1/2)\left[\frac{1}{1 + Y(2A + z)/k} + \frac{1}{1 + Y(2A + z)/k}\right].$$

If we write $B = (2A + z)Y/k$, $x = zY/k$, and substitute these into the equations for q and q', we obtain

$$q' = 1/(1 + B);$$

$$q = (1/2)\left[\frac{1}{1 + B + x} + \frac{1}{1 + B - x}\right] = \frac{1 + B}{(1 + B)^2 - x^2}.$$

Thus q exceeds q': an "optimum" family contributes more than an average family, no matter what the average family composition of the population.

Problem

In 1956, 1,820,700 white boys, 1,724,000 white girls, 312,800 Negro boys, and 304,900 Negro girls were born in this country. Of these, 68,800 white boys, 47,450 white girls, 19,400 Negro boys, and 14,900 Negro girls died before their fourth birthday. Assume that any child dying before age four costs its parents nothing, and that any child living past this age matures.

1. Are the sex ratios of the two sexes adapted to present-day death rates?

2. For each race, calculate the multiple of the present-day death rate to which the sex ratio is adapted. (Data from Kolman.)

Solution: The ratio of male to female births should be the ratio of the prospective cost of a female birth to that of a male. Since 3.78% of the white boys and 2.75% of the white girls die before their fourth birthday, the neonate sex ratio among whites should be 97.25/96.22, or 101 boys to 100 girls. In fact, it is 1,820,700:1,724,000, or 105.6:100.

Suppose now the sex ratio is adjusted to a death rate k times higher than the present. How large is k? k is determined by the equation

$$\frac{100 - 2.75k}{100 - 3.78k} = 1.056.$$

Solving, we find $k = 4.5$: among whites, the neonate sex ratio is adapted to death rates of 17.0% among boys and 12.4% among girls.

Among the Negroes, 6.2% of the boys and 4.9% of the girls died before their fourth birthday. The appropriate ratio is 101.4 male births for every 100 females. The actual sex ratio is 102.6 : 100, which is adapted to a death rate 1.8 times higher than the present, when 11% of the boys and 8.7% of the girls die before their fourth birthday.

Bibliographical Notes

The discussion of sex ratio is drawn primarily from R. H. MacArthur, "Ecological Consequences of Natural Selection," pp. 388-397 in T. H. Waterman and H. J. Morowitz, eds., *Theoretical and Mathematical Biology*, Blaisdell, 1965.

Levins was the first to convince biologists of the usefulness of budget lines (the boundaries of his fitness sets) and indifference curves (the contours of his adaptive function): see his paper "Theory of Fitness in a Heterogenous Environment I: The Fitness Set and Adaptive Function," pp. 361-373 in the *American Naturalist*, vol. 96, 1962.

Although it is convenient to assume that successive generations are distinct, that one generation dies before any of the next achieve sexual maturity, MacArthur's conclusions apply equally to populations where individuals are continually dying and being born: see E. Leigh, "Sex Ratio and Differential Mortality between the Sexes," pp. 205-210 of the *American Naturalist*, vol. 104, 1970.

CHAPTER 4

How Do Shells Grow?

WHAT LIMITS possible strategies? For sex ratio, we considered the energy expenditure required to bring up children, and asked how much the parents could afford. The whole bent of our biological training impresses on us the importance of energy: elementary textbooks are grossly enlarged and distorted by the need to discuss it. Indeed, those chapters of elementary biology most remote from the realm of sense and feeling, the endless disquisitions on elementary chemistry, the entrancing yet seemingly unreal discussions of photosynthetic processes, of Krebs cycles and the like, are all part of the earnest attempt to confer some understanding on the student of the use and transfer of energy.

The importance of energy is impressed on us simply by growing up in today's world. The history of industrial development bespeaks its importance, for the development of convenient and compact sources of inanimate power always paces the first stages of industrialization. The social importance of its availability is immortalized in the grand title of the paperback, *Energy and Society,* which you may have noticed in your bookstore. But as industry becomes more complex, directing the industrial process becomes every bit as serious a problem as supplying energy for it. The industrialist must increasingly concern himself with the transfer of *information.* Thus a new branch of engineering, concerned with communications, has grown up to take its place beside its older brother, concerned with power. And, just as the concept of energy was once so much the rage, now the concept of information is in fashion.

An organism also has problems of information transfer. Indeed, the great revolution in biology this century has hinged upon the discovery of the means of heredity, the mechanism of information transfer in organisms. Elementary biology tells us how characteristics are conveyed from parents to offspring primarily as sets of instructions encoded in the base-arrangements of the double helix of desoxyribose nucleic acid, DNA. The blueprint of the form, structure, and other characteristics of an organism are thus contained in the DNA of the fertilized egg; half of this blueprint was contributed by a small sperm-cell. A sperm-cell can hold only a limited amount of DNA: hence the number and complexity of the instructions it can carry are limited. They could not possibly specify the building of a man molecule by molecule; they probably could not even specify the arrangement and interconnections of the many kinds of cells that make up a man. A DNA blueprint, however, differs from that of an architect: the architect specifies a finished structure, while the DNA specifies processes: to be precise, a set of interlocking chemical reactions which, under the right conditions, will make an organism. The plans for a building are not necessarily altered by accidents occurring during construction: work can be modified or begun anew to construct it as shown in the blueprint. If, however, a growing organism is subjected to a heatshock that derails its developmental processes, a perhaps inviable deviant may result: there is no structural blueprint the derailed developmental process can consult to set itself right again. The complexity of a finished structure does not imply equal complexity in the processes forming it; thus the complexity of humans does not prove they were not constructed from DNA blueprints. The small size of sperms does, however, imply some limit to the complexity of developmental process.

Why not make sperms a little larger? Probably because it is expensive to read information from DNA. Let us resort once again to analogy. Humans may be considered the DNA of industry, the ultimate specifiers of its processes. One characteristic of modern times is that the most economical industrial processes require the least human interference. There is no shortage of available personnel, as the strikes associated with automation testify: it just happens that human beings are more expensive and make more mistakes than machines. Control by the DNA of a step in a chemical process requires the manufacture of an enzyme; it may also require a governor to prevent excess or deficiency of that enzyme. This machinery may be both expensive and delicate, and it may pay the organism to be economical of it.

Consider a shelled animal. The pattern of shell growth will determine the shell's shape. Can different manifestations of a simple growth-pattern, economical of information, yield shell forms appropriate to different ways of life? A shelly exoskeleton imposes tremendous limitations on growth: either the shelly covering must be shed periodically and a new and larger one grown in its place, with all the risk this entails, or the shape of the animal must conform to a shell which can only be added to along its margins, and which therefore must be essentially conical. The first is the strategy of the arthropods, the second that of the mollusks.

Perhaps the simplest sort of developmental pattern is one

which does not change as the animal grows. Consider an animal whose shell is a permanent fixture, and assume that both shell and animal preserve their shape as they grow, merely increasing in size, as under a magnifying glass. The shell's new growth will then always stand in the same relation to the animal, so that such shells are indeed the outcome of a fixed developmental pattern. What shell forms can be generated this way, and what are they good for?

One example is a bilaterally symmetrical animal which drags his shell after him. If the shell preserves its shape as it grows, it will be conical, like a dunce cap. Hundreds of millions of years ago there lived cephalopods, relatives of the modern-day octopus, squid, and nautilus, which possessed such shells. Once all these straight-shelled cephalopods were thought to be related, and they were all given the name *Orthoceras*, but paleontologists have since learned that possession in common of such a simple shell form may be a matter of coincidence, and by no means implies recent common ancestry.

The cephalopod never filled his conical shell entirely, and he squeezed out of its hind end as he grew. The resulting empty space was apparently filled with gas, as in the modern nautilus, except for a slender tubular organ tapering back to the very end of the shell (Fig. 4-1). Thus shell and animal became buoyant and free-floating. To seal in the gas, the animal formed a partition or *septum* behind its body every once in a while. Some animals formed their septa in geometric progression, so that each chamber was similar to the one before and the shell as a whole preserved its form. The tube extending back to the end of the shell secreted water or laid down lime to balance the shell and keep it horizontal: the weight and distribution of the secreted lime would be such that the animal could either crawl or swim, as conditions demanded.

A swimming cephalopod with a long, slender shell would be streamlined and speedy, one of the fastest animals in the

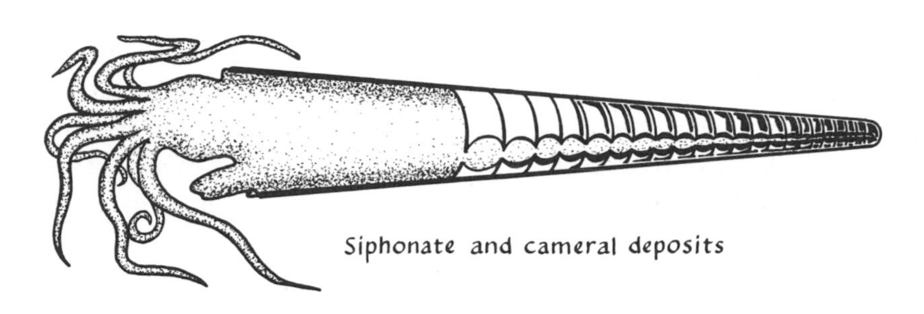

Fig. 4-1. A straight-cone cephalopod. The stippled area represents the animal: notice the siphuncle, through which the animal extends back to the hinder end of its shell. The blackening under the siphuncle and toward the rear of the shell represents cameral deposits, intended to preserve the shell's balance. (After Fig. 9-4, no. 3, from Raymond C. Moore, C. G. Lalicker, and A. G. Fischer, *Invertebrate Fossils*. Copyright 1952 by McGraw-Hill Book Company. Used with permission of McGraw-Hill Book Company.)

whole ocean. We may presume that, like their modern descendants, the squid, they moved by jet propulsion, squirting water with great vigor when circumstances required. However, such long shells would not be very maneuverable.

The septa strengthened the shell, just as a bamboo stem is strengthened by its rings. Still, such a long, slender object would be easily broken. Were it knocked in the middle, the bending stress would be proportional to weight times length (think of the shell as a lever about the breaking point), while its power to resist this stress would be proportional to its cross-sectional area. Therefore, among geometrically similar *Orthoceras*, fragility increases as the square of the length.

Were we to wind up the *Orthoceras* into a coil, making a nautilus or ammonite of it, we would have a far stronger shell.

Not only would we reduce its maximum dimension, but to break it across the middle one would have to break, not just a single tube, but all the whorls together. Around reefs and in turbulent places, therefore, one would expect selection to favor coiled forms over straight. The compact ammonite is also more maneuverable, and the intrinsically greater strength of its form sometimes allows a thinner, lighter shell. The contrast between the maneuverable coiled form and the streamlined straight one is like that between the Volkswagen and the Lincoln Continental: each is useful, but for a different purpose.

What must a coiled shell look like if it retains its shape as it grows? To ensure that it tends to move in a straight line, the cephalopod, like the jet plane, will be bilaterally symmetrical. Let r be the distance, in millimeters, from the axis of coiling to the outermost point on the lip of the shell: we call r the shell's *radius*. The radius is a function $r(x)$ of the number x of revolutions the shell has made about its axis, or, in the shell collector's phrase, its number of whorls. The graph of $r(x)$ in polar co-ordinates is the spiral curve we would see if we sawed our shell in half along its plane of symmetry (Fig. 4-2). If the shell preserves its form as it grows, growth is equivalent to a simple magnification of the shell form (Fig. 4-3). The shell will be magnified in equal proportions as it grows through equal angles. Thus, if the shell has a one-millimeter radius to begin with, and a radius of b millimeters one revolution later, it will have a radius of b^n after n revolutions. Moreover, $1/m$ of a revolution magnifies the spiral by a factor of $\sqrt[m]{b}$, or $b^{1/m}$. The shell's radius after n/m revolutions is $b^{n/m}$: the equation for the spiral curve followed by the outer lip is thus $r(x) = b^x$. The number of revolutions of growth is the logarithm, to the base b, of the shell's radius, so the curve is called the logarithmic spiral.

However, the form of a shell must be represented by a

Fig. 4-2. An ammonite sawed in half to reveal the logarithmic spiral formed by its shell. (Photograph by S. Rawson.)

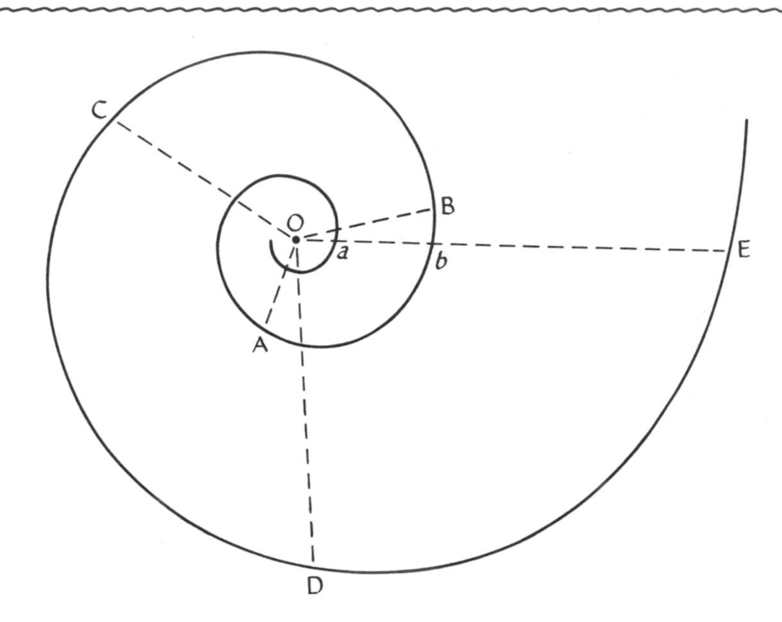

Fig. 4-3. A logarithmic spiral, swept out by a radius vector which expands as it revolves. *OA, OB, OC, OD,* and *OE* represent successive positions of the radius vector. The lengths of *Oa, Ob,* and *OE* are in geometric progression.

surface, not merely a curve. As we imagined the curve of the last paragraph swept out by a point attached to a radius vector which expanded as it revolved, so now our surface is swept out by a figure representing the cross-section of a whorl attached to the end of our expanding radius vector (Fig. 4-4). This figure is called the shell's *generating curve*: if the radius vector and its attached figure expand in proportion as they revolve about the shell's axis, the curve will "generate" a surface which preserves its shape as it grows. The cone is only one of a family of such shell forms, the one where the shell grows uniformly around the lip. There is no a priori reason why the rate of shell growth on the under and upper sides of the animal should be the same, as these sides are unlike in other respects, and, indeed, the first cephalopod shell was slightly curved.

Fig. 4-4. A wentletrap. Imagine a line connecting the shell's apex with its aperture (mouth). The aperture is the final position of the "generating curve": we imagine that the line and attached curve expand geometrically as they revolve around the shell's axis, the curve "generating" the shell in the process. The ribs mark successive positions of the generating curve.

The direction of growth of a logarithmic spiral always makes the same angle with the radius vector (see Appendix), so that the length of the spiral is proportional to the radius. The width of a shell's aperture is proportional to the shell's radius, and thus to the length of the unwound shell. If we could straighten out a spiral shell we would obtain a cone: that is why we thought of *Nautilus* as a coiled *Orthoceras*. The outer and inner edges of our coiled cone are described (in side view) by the spirals $r = b^x$ and $r' = Ab^x$ respectively (Fig. 4-5): the rate of shell deposition on the inner edge of the shell is reduced by a factor A relative to the outer. If $Ab^x < b^{x-1}$, the outer whorl will overlap its neighbor to the inside: the conditions "A is greater than, equal to, or less than $1/b$" govern whether the

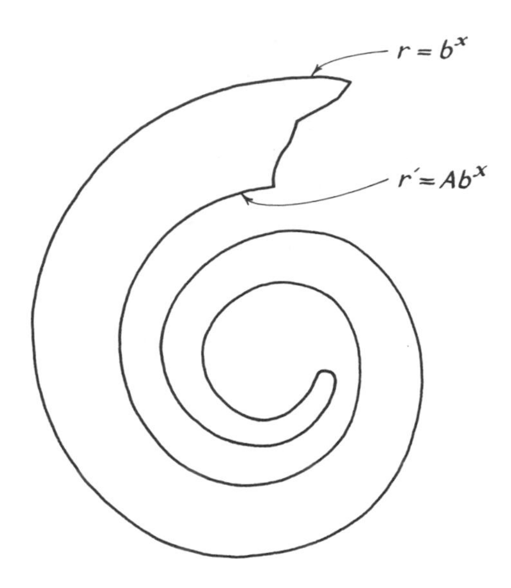

Fig. 4-5. The coiled cone of a snail shell. The spirals $r = b^x$ and $r' = Ab^x$ describe respectively the outer and inner edges of the cone. Notice that if A were less than $1/b$, successive whorls would overlap.

shell is an open discoidal spiral, whether the whorls just touch, or whether they actually overlap. Thus the factors A and b describe the shell as it appears from side view. If $A = 0$, the outer whorl overlaps all the rest, which condition is that of greatest strength; however, the space occupied by the preceding whorls gives the aperture a horseshoe-shaped cross-section. In *Nautilus*, the aperture is of a relatively convenient shape, because the whorls expand at such a rate that the arms of the horseshoe are no great part of the aperture as a whole (Fig. 4-6); but if the shell were wound very tightly, the shape of the aperture might seriously restrict the ways of life open to the occupant.

Now for crawling snails. When *Nautilus* crawls, its shell is unstable, and must be buoyed up. Were the shell wider and

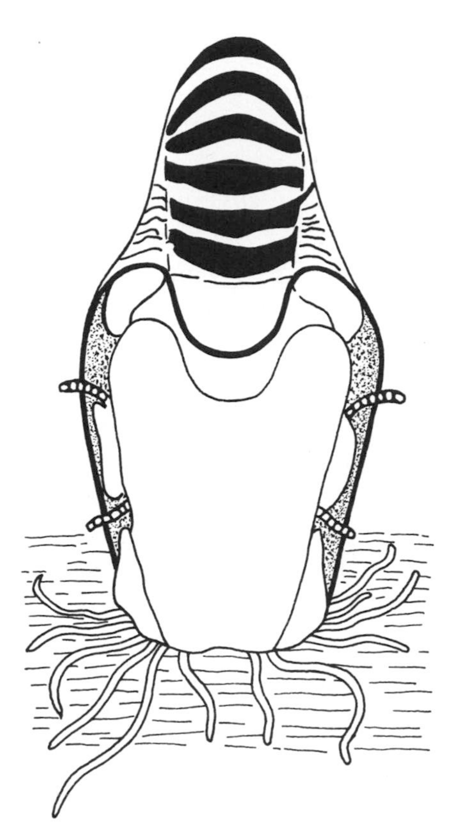

Fig. 4-6. A crawling nautilus: the shell is buoyed up by the gas it contains. Note the roughly horseshoe shape of the aperture. (After Fig. 442-A, p. 649, of L. A. Borradaile et al., *The Invertebrata: A Manual for the Use of Students*, 4th edition, revised by G. A. Kerkut, 1963, with permission of the Cambridge University Press, Cambridge, England.)

more tightly wound, it would sit lower on the ground and be less likely to flop over, but then it would be poorly streamlined, and the shape of the aperture would be restrictive. To be suitable for crawling, a bilaterally symmetric shell would have to be wide and rapidly expanding; an extreme example is the coolie-hat shape assumed by the (secondarily symmetric) limpets of today. The bilaterally symmetric ancestors of today's snails had shells only slightly more curved, still essentially cap-shaped. The same is true of *Neopilina*, a "living fossil" from the deep sea, the only snail alive which preserves the ancestral bilateral symmetry. But the occupant can hardly retreat into and close off a cap; limpets now substitute the rock they crawl on for an operculum*, closing their shell against it when danger threatens, so that poets marvel at their tenacity. But that response is suitable only when one is crawling upon a rock: today's limpets live on wave-beaten cliffs and rocky coasts, not on soft-bottomed flats.

If a snail shell is to taper slowly enough that the animal can retreat into it, then the cone should be wound along a spindle rather than in a plane coil. This provides a more compact shell without restricting the shape of the aperture (see Fig. 4-4), and the shell gains strength from the way the whorls overlap. This also allows the cross-sectional area of the shell in the direction of forward motion to be reduced without reducing the shell's volume. In short, "conispiral" coiling permits a greater and more satisfactory variety of responses to the demands of stability, streamlining, shell strength, animal shape, etc. which different ways of life may impose. Today, most crawling snails have conispiral shells; even the limpets, for all their symmetry, are conispiral in their early stages. Moreover, every univalve mollusk crawling today, with the exception of *Neopilina*, is descended from conispiral stock: it is as if the evolutionary plasticity inherent in the conispiral shell form were necessary for a stock of crawling snails to survive the revolutions of the ages.

How does an animal grow a conispiral shell? Formally speaking, a conispiral shell is generated by the cross-sectional

---

* The operculum is the calcareous or horny "door" with which many snails close off their aperture when they retreat into their shell.

figure attached to a radius vector which is declined relative to the axis about which it revolves. The expanding radius vector now sweeps out the surface of a cone, rather than a plane surface. The declination destroys the bilateral symmetry of the shell: the conispiral is the product of an asymmetric growth process. Otherwise, nothing is changed: the radius vector and generating curve expand as they revolve, in the same proportion each revolution. The shell still preserves its shape as it grows, as we may learn from certain turbans and moon-shells. As the aperture keeps its shape, so must the operculum also; and these opercula, which grow at one end only, preserve traces of logarithmic spirals (Fig. 4-7).

Fig. 4-7. *Natica* with operculum. *Natica* preserves its shape as it grows, and so must its operculum, which therefore forms a logarithmic spiral. (Photograph by S. Rawson.)

The conispiral embraces the most general class of shells which preserve their shape as they grow. Such spirals are not restricted to shells: Thompson tells us how claws and teeth, horns and tusks, are often curved thus, and how we may expect to see this curve in any rigid structure which grows at one end only, according to an unchanging pattern where each stage of growth remains an integral part of the whole. This spiral, however, is characteristic only of the rigid and the dead: we are not to seek it among the curves of a fish, or the curls of a girl's hair.

Appendix – The Logarithmic as an Equi-Angular Spiral

The logarithmic spiral may be derived from the property that its direction of growth always forms the same angle with the radius vector. We may analyze (Fig. 4-8) a fraction dx of a revolution's growth into a radial component dr and a tangential component $2\pi r dx$ (the 2π enters because a full revolution of tangential growth would be $2\pi r$) perpendicular to the radius. If the new growth always makes the same angle with the radius vector, then the ratio of the two components of growth must be constant. Calling this constant c, we may write

$$dr/2\pi r dx = c;$$

$$d \log r = dr/r = 2\pi c dx;$$

$$r = e^{2\pi x c},$$

where $e = 2.718\ldots$, the base of the Napierian logarithms. If we set the Napierian logarithm of b equal to $2\pi c$, our equation becomes

$$r(x) = b^x,$$

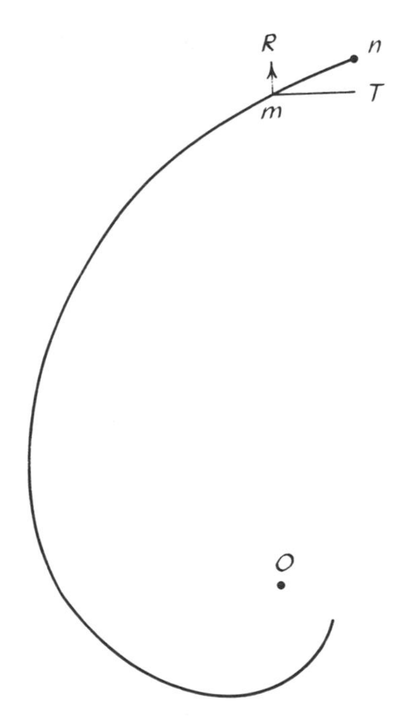

Fig. 4-8. Dynamical aspect of the equi-angular spiral. Its growth from *m* to the neighboring point *n* may be analyzed into a radial component (growth in the direction *R*) and a tangential component due to rotation of the radius vector *OR* (growth in the direction *T*). In an equi-angular spiral, the ratio of radial to tangential growth does not change. (After Fig. 352 of D'Arcy Thompson, *Growth and Form*, 2nd edition, 1942, vol. 2, p. 756, with permission of the Cambridge University Press, Cambridge, England.)

the equation of the logarithmic spiral. This derivation shows why so many call this curve the equi-angular spiral, and demonstrates in a very concrete way why any shell produced by an unchanging pattern of growth must assume this spiral form.

Problem

Is the shell in Fig. 4-2 a logarithmic spiral? Are the septa so placed that the shell preserves its form as it grows?

Solution: Lay a ruler from the "origin" of the shell to its edge, and measure the distances from the origin to the places where the ruler intersects the spiral. To verify that these distances form a geometric progression, find whether their logarithms are in arithmetic progression.

The number of septa per revolution declines as the shell grows, so the shell's form does change in this respect.

Bibliographical Notes

An interesting discussion of "communications engineering" is given in the first 44 pages of N. Wiener, *Cybernetics*, M.I.T. Press (first edition 1949, second 1961). It is a curiously difficult book. Its author often seems unsure of himself in subtle ways, as if he were not quite sure whether he was a quack; in fact, he was one of the great mathematicians of the century, and also the first to popularize the ideas of communication and control which biologists now find so commonplace. Although the book has many pages of Fourier integrals, it also has many pages which will profit the non-mathematical reader.

Small viruses have coats composed of a multitude of identical subunits because they cannot hold enough DNA to specify a coat consisting of a single large molecule: see F. Crick and J. Watson, "The Structure of Small Viruses," pp. 473-475 in *Nature*, vol. 177, 1956, and "Virus Structure:

General Principles," pp. 5-13 in G. Wolstenholme and E. Millar, eds., *CIBA Foundation Symposium on the Nature of Viruses*, Churchill, 1957. The subunits, of course, are constructed in such a way that they will assemble *automatically* into protein coats, as Caspar and Klug so beautifully describe in "Physical Principles in the Construction of Regular Viruses," pp. 1-22, of *Cold Spring Harbor Symposia on Quantitative Biology*, vol. 27, 1962. Ideas concerning the economical use of information undoubtedly have a very wide application; see, for example, M. Steinberg, "Does Differential Adhesion Govern Self-Assembly Processes in Histogenesis? Equilibrium Configurations and the Emergence of a Hierarchy among Populations of Embryonic Cells," pp. 395-434 in *Journal of Experimental Zoology*, vol. 173, 1970.

The discussion of logarithmic spirals is based on the eleventh chapter of D'Arcy Thompson's book, *Growth and Form* (Cambridge University Press, first edition 1917, second 1942). This is one of the classics of biology, an attempt to elucidate morphology in terms of geometry and physics. Thompson was interested in biological mechanisms; thus he disliked genetics, where chromosomes were supposed to exercise profound influence by invisible means (genetics is so mechanist nowadays that it is hard to remember how occult it seemed when he wrote), and he distrusted the theory of natural selection, which seemed to be replacing natural theology as an excuse for laziness and intellectual dishonesty. Isolated from the "mainstream of biology," he brought to the subject a completely fresh viewpoint, which sometimes led him into disastrous error but more often was strangely revealing. An abridgement, with comments on the present-day status of Thompson's theories, has been published by J. T. Bonner (Cambridge University Press, 1961).

Further discussions of shell form may be found in D. Raup, "Geometric Analysis of Shell Coiling: General Problems," pp. 1178-1190 of the *Journal of Paleontology*, vol. 40, 1966, and "Geometrical Analysis of Shell Coiling: Coiling in Ammonoids," pp. 43-65 of the *Journal of Paleontology*, vol. 41, 1967.

A readable account of the mollusks is given in the early chapters of R. Tucker Abbott's *American Seashells,* Van Nostrand, 1954.

CHAPTER 5

The Evolution of Snail and Squid

IN THE last section we discussed some factors influencing the evolution of shell form, and showed how these could explain certain evolutionary trends among mollusks. Now we shall arrange these explanations side by side to obtain an overview of the major features of evolution among gastropods and cephalopods. Some readers may need to refer to the glossary of zoological terms and geological period names given in the appendices to this chapter.

What were the general features of the stock from which mollusks derived? The presence of a similar "trochophore larva" stage among some annelids, arthropods, and mollusks suggests that these three phyla derived from a common root. The common ancestry of arthropod and annelid is suggested by the early trilobites, which seem hardly more than baroquely elaborated, shell-bearing segmented worms; the common ancestry is likewise suggested by *Peripatus*, that marvellous intermediate between arthropod and annelid, whose near relative is so splendidly preserved in a Cambrian formation. The kinship between mollusks and annelids is suggested by *Neopilina* (Fig. 5-1), whose shell so resembles those of its presumptive ancestors in the Cambrian. The anatomy of this snail displays a "metameric" structure, tending to repeat itself from front to rear, suggesting the segmentation of the annelids. We therefore have reason to believe that the ancestors of the mollusks were bilaterally symmetrical and possessed a segmented or metameric structure. Originally the segments may all have been alike, but the problems of sensation in an animal which moves *forward* would eventually have led to the development of special organs in the front segments. The front segments of early trilobites, for example, are markedly changed, while the rear ones closely resemble those of the worms their ancestors. Our ancestral worm would also reflect the difference between up and down, but, since there is no inherent difference between the world on an animal's right and that on its left, we would expect bilateral symmetry to persist until the animal's internal architecture required a departure from it. Just before the beginning of the Cambrian, ancestral mollusks very likely retained considerable metamerism, and a bilateral symmetry reflecting a nearly absolute symmetry of developmental process.

The first shelled ancestors of the gastropods were the bilaterally symmetric monoplacophorans, which appeared in the Cambrian. Some of these had cap-shaped shells; others had coiled shells they could retreat into. To judge by their modern descendant *Neopilina*, they were crawlers rather than swimmers; like the chambered nautilus, they wore their shell so that the coil faced frontward and the growing lip behind (Fig. 5-2). The shell was a protection from predators and probably also a shield from ultraviolet radiation, little of which was filtered by the atmosphere in those days.

Gastropods probably arose from the *Monoplacophora* by way of the bellerophontids, which had coiled, bilaterally symmetric shells, notched in the center of the outer lip. This notch often became a long slit in the median plane, traces of

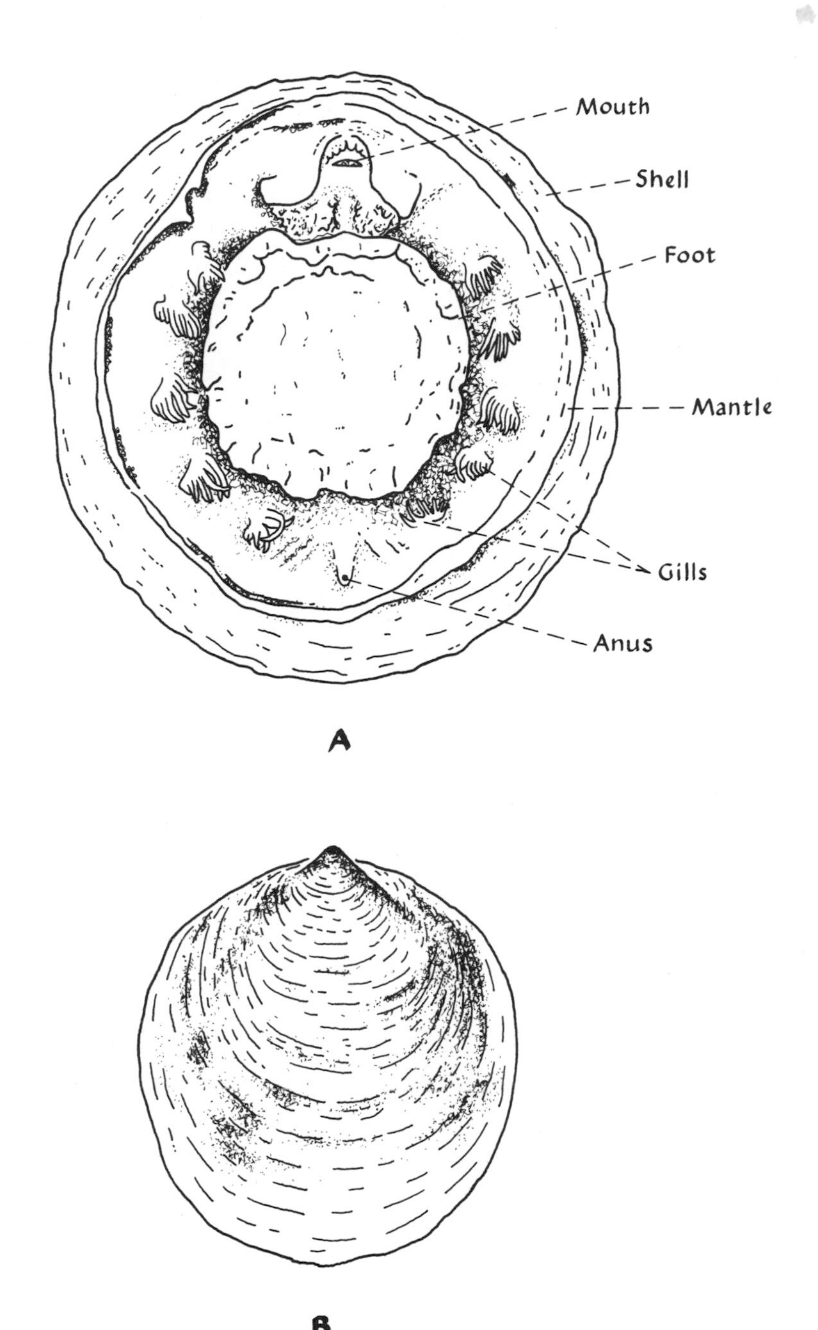

Fig. 5-1. Neopilina. (Adapted from Lemche and Wingstrand in Robert Barnes, *Invertebrate Zoology*, 2nd edition, W. B. Saunders Company, Philadelphia, 1968.)

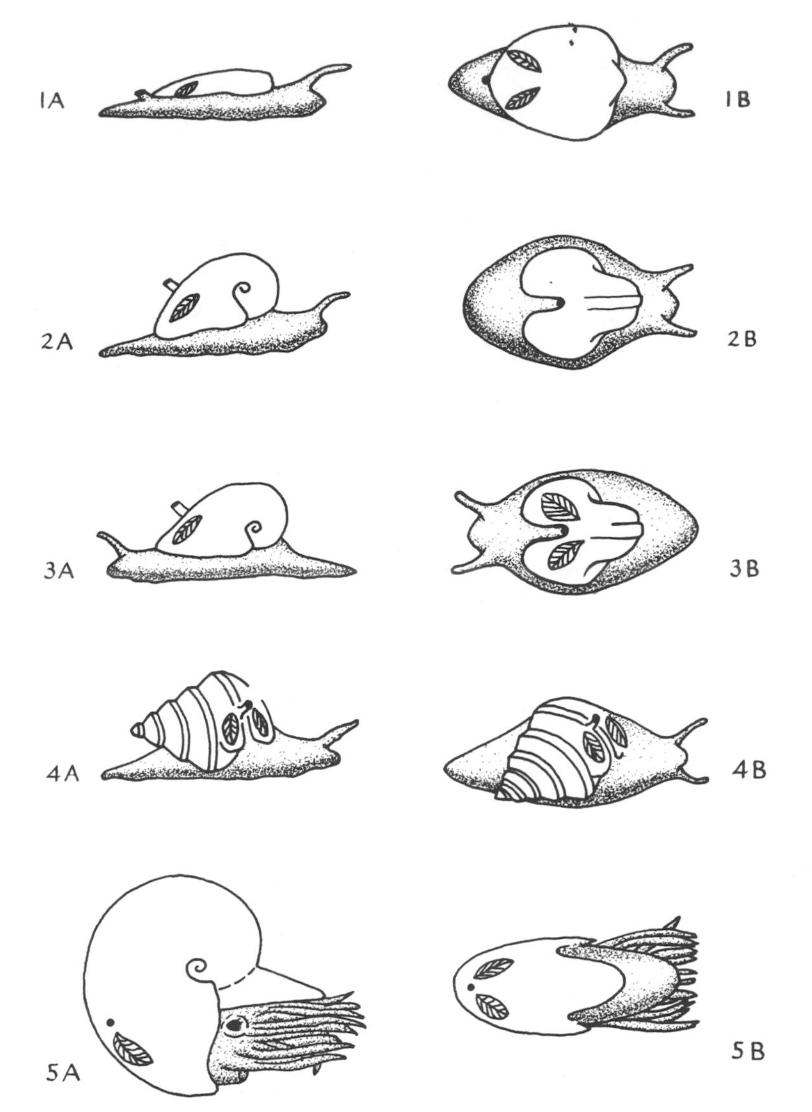

Fig. 5-2. Orientation of shells on torted and untorted animals (side and top views): 1A and 1B, *Monoplacophora*; 2A and 2B, hypothetical untorted *Bellerophon*, with coil facing frontward; 3A and 3B, *Bellerophon*; 4A and 4B, *Pleurotomaria*; 5A and 5B, *Nautilus*. *g* and *a* denote gills and anus, respectively. (After Fig. 8-14 of *Invertebrate Fossils*, by Moore, Lalicker, and Fischer. Copyright 1952 by McGraw-Hill Book Company. Used with permission of McGraw-Hill Book Company.)

whose former positions left a depressed scar, the *selenizone*, in the shell. Most bellerophontids could retreat into their shells for shelter. The bellerophontids gave rise in turn to two conispiral groups, the macluritids and the pleurotomariids. These groups differ in their manner of coiling (Fig. 5-3); the macluritids, moreover, had only a rather shallow notch in the lip, leaving a scar in the form of a raised spiral "keel" on the edge of the shell, while the pleurotomariids possessed a well developed slit and selenizone (Fig. 5-4).

The macluritids and their descendants died out in the Mesozoic, but pleurotomariids survive today in the deep sea. Related forms, some of which possess slit conispiral shells as larvae, preserve traces of the slit in the adult shell, as the hole atop a keyhole limpet, or the row of holes in the abalone. Modern pleurotomariids and their relatives possess paired gills and other traces of bilateral symmetry which are lost in more "advanced" forms.

In the pleurotomariids, the position of the shell on the crawling animal is such that the growing lip faces forward, rather than behind as in *Neopilina* (see Fig. 5-2). This improves streamlining and ensures that the snail exerts its pull on a part of the shell forward of its center of gravity, so that the shell rests more stably upon the moving animal. This reversal of the shell's orientation on the animal occurs during larval development; the folding or twisting of the larva which brings this about is called *torsion*.

The pleurotomariids, possessing the twin properties of torsion and conispiral coiling, gave rise to all the surviving gastropods. So important are these two properties that nowadays every crawling mollusk save *Neopilina* and the chitons (whose shells are essentially cap-shaped: Fig. 5-5) is descended from conispiral stock and undergoes torsion during larval development. Gastropod evolution since the origin of the pleurotomariids may be interpreted as the exploitation of the ways of life opened to crawling snails by the development of these two

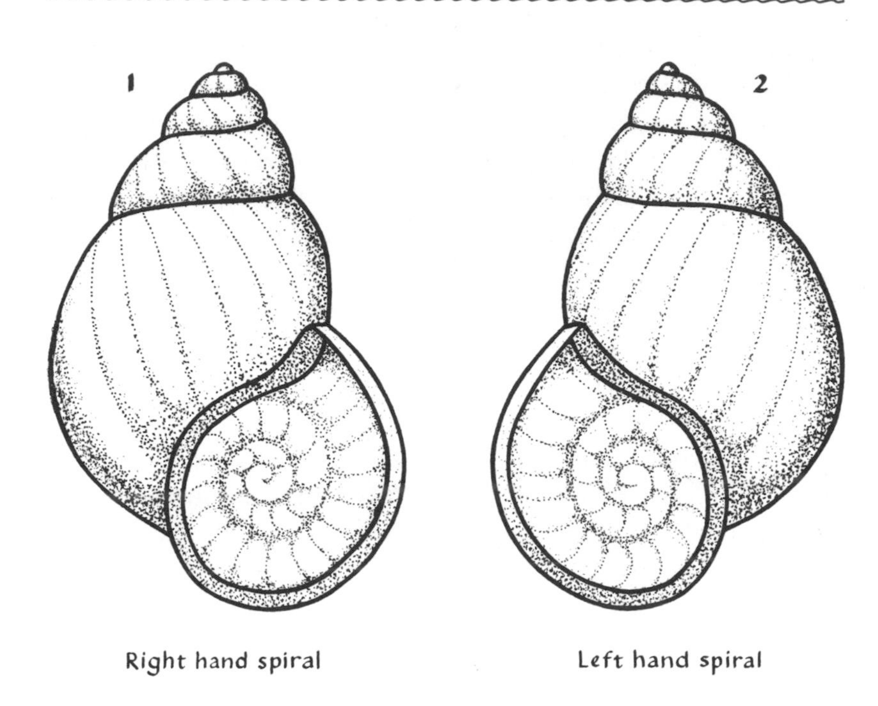

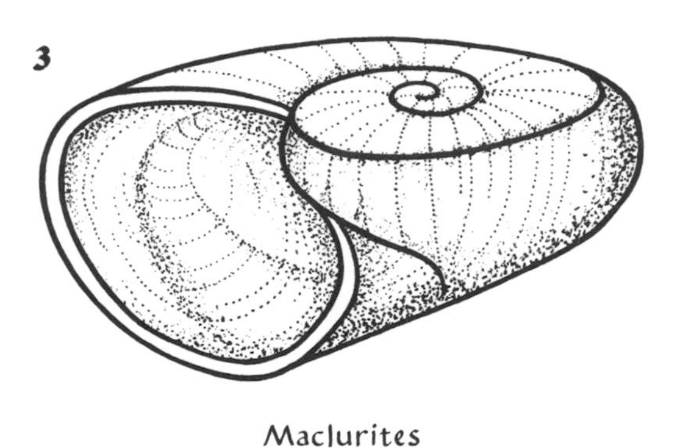

Fig. 5-3. Most gastropod shells are right-handed, or clockwise, spirals (1), but some are left-handed (2). Many gastropods are equipped with an operculum, which closes the aperture when the animal withdraws into its shell: if it shows spiral structure, its spiral is opposite to that

Fig. 5-4. Pleurotomaria. (Photograph by S. Rawson.)

of the shell. Gastropods ordinarily are illustrated with the apex of the shell pointing upward. When the Ordovician *Maclurites* (3) is oriented in what appears to be the conventional way, it seems left-handed. The counterclockwise structure of its operculum demonstrates, however, that this shell is actually a right-handed coil with its true apex hidden by the enlarging whorls: this specimen is illustrated wrong side up. (From Fig. 183a of *The Course of Evolution*, by J. M. Weller. Copyright 1969 by McGraw-Hill, Inc. Used with permission of McGraw-Hill Book Company.)

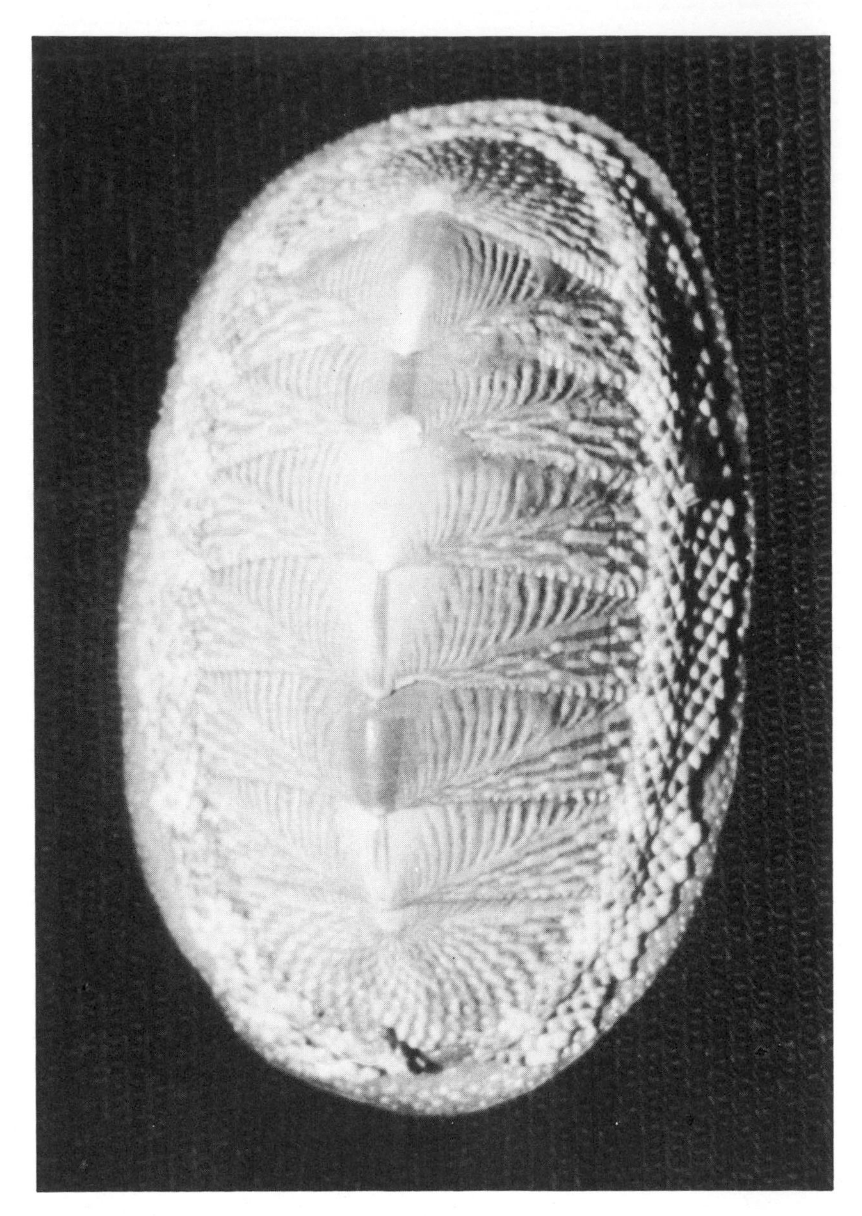

Fig. 5-5. A chiton. (Photograph by S. Rawson.)

properties. Fossil strata record the adjustments of shell shape to differing balances of demands for streamlining, stability, and strength. There are many variations on this theme, from limpets streamlined to take the full force of waves on a rocky coast to the heavy shells of Panamanian *Muricanthus*, prickly and spiny like the fruits of cacti so that they might not be eaten (Fig. 5-6).

Fig. 5-6. Muricanthus. (Photograph by S. Rawson.)

Both torsion and conispiral coiling result from processes which are not bilaterally symmetric. Such an asymmetry was probably hard to come by in a stock as fundamentally symmetric as the ancestral mollusks: perhaps this is why conispiral coiling only evolved twice during the hundred million years of the Cambrian. Textbooks tell us torsion was a unique event in molluscan history: it evolved only in the bellerophontids, from which it was inherited by both macluritids and pleurotomariids.

Why should torsion have occurred in the bellerophontids? The fact that ancestors and descendants of the bellerophontids crawled suggests that the bellerophontids did too; surely a crawling habit would have enhanced the advantages of torsion. On the other hand, if conispiral shells are so well suited to crawlers, why did the bellerophontids persist for three hundred million years in the face of conispiral competition? Perhaps they retained their symmetry because they swam as well as crawled, as some sea slugs do today.

Now for the cephalopods. They first appeared in the Cambrian with conical, slightly curved, septate shells which quite clearly served as shield and float. These shells were curved in one plane, reflecting the bilateral symmetry of their owners. The successive septa and the resulting repetitive structure of the siphuncle seem to reflect the metameric structure of ancient ancestors, as if the posterior part of a cephalopod, like that of a trilobite, most nearly resembled the corresponding part of the worm from which it derived. We have seen how the cephalopods produced both straight-cone and coiled forms, and how some of the straight-cone forms were crawlers, and others speedy swimmers, while the coiled forms were slower, more maneuverable swimmers, with shells either lighter or stronger than their straight-cone counterparts. The straight-cone crawlers were succeeded by gastropods, whose asymmetry permitted shells better suited to crawling. The straight-cone swimmers were presumably displaced by the even speedier squid, which itself derived from one of them. Perhaps improved musculature and nervous co-ordination developed step by step until the swimming cephalopod could dispense with his float, even find it an impediment, like the person who has learned to swim. The coiled forms lasted longer, but fifty million years ago they were nearly all extinct, perhaps again displaced by shell-less forms.

Cephalopods were more bound than their gastropod relatives by the ancestral tendency to bilateral symmetry. Only once, in the Devonian, did the nautiloid cephalopods give rise to conispiral stock; the ammonites evolved conispirals in the Triassic and Cretaceous. It would appear that the ancient cephalopods, like their modern shell-less descendants, tended to swim rather than crawl, and that as time passed this tendency became stronger. At first, it is difficult to see why this should be. There were crawling cephalopods in the Ordovician, and surely a crawler with arms like an octopus could do many things a snail could not. The most likely answer is the limitation of symmetry: it must be very difficult to obtain an asymmetric pattern of development from one which is basically symmetric. The asymmetric developments of torsion and conispiral coiling did occur in some Cambrian snails, but such mutations would presumably be so rare, and involve such a wrench of developmental process, that they would hardly be favored against established competition. Such are the limitations of symmetry that Mother Nature seems to have found it easier to teach shrimps to *wear* cast-off conispiral shells and become hermit crabs, rather than to teach small cephalopods to grow them.

Appendix 1 – Classification

Living things are classified according to a hierarchical scheme. The largest divisions are the kingdoms, animal and plant. These are subdivided into phyla: an example of a phylum is the *Chordata*, the animals which possess a backbone or notochord at some stage in their life cycle. The phylum *Chordata* is subdivided in turn into classes: fishes, birds, mammals, etc. Further subdivisions include orders (such as the primates, embracing lemurs, monkeys, and man), families (the great apes form a family; the Old World Monkeys another), genera, and species, which last are roughly the kinds of the Bible. In the preceding chapter, we mentioned the following phyla, classes, and subclasses of invertebrates:

Phylum *Annelida*: segmented worms, such as earthworms and leeches.

Phylum *Arthropoda*: "jointed-legged animals" which possess a chitinous exoskeleton: crabs, insects, spiders, millipedes, scorpions, and the like. The exoskeleton must be shed periodically to form a new and larger one. Like the *Annelida*, the arthropods began as segmented creatures, each segment possessing a pair of appendages. These appendages undoubtedly all began as feet, but some have since been transformed into claws, antennae, jaws, etc.

Subphylum *Onychophora*: *Peripatus*. Very like annelids, with internal segmentation and a soft skin. The circulatory and respiratory systems are quite arthropod-like.

Subphylum *Arthropoda*: more "usual" arthropods.

Class *Trilobita*: trilobites. Flattened animals, which crawled along the sea-bottom. Their exoskeletons were sometimes quite hard. The body was divided into three parts, a head end or "cephalon" consisting of six fused segments, a many-segmented thorax, and a tail-piece. Longitudinally, the body appears formed of a central axis with a lateral lobe on each side, whence the name "trilobite."

(Class *Crustacea*: crabs, shrimps, water-fleas.)

(Class *Chelicerata*: spiders and their allies.)

(Class *Insecta*: insects, millipedes.)

Phylum *Mollusca*: snails, clams, and the like. Shell, when present, a permanent fixture.

Class *Monoplacophora*: bilaterally symmetric "snails."

Class *Gastropoda*: snails and slugs. All gastropods undergo torsion and form a conispiral shell at some stage in their life cycle.

Class *Amphineura*: chitons. Shell, when present, consists of eight plates surrounded by a girdle. Bilaterally symmetric.

Class *Cephalopoda*: squids, octopi, nautilus.

Subclass *Nautiloida*: shelled cephalopods with simple septa.

Subclass *Ammonoida*: shelled cephalopods. In contrast to the nautiloids, the shells are often ornately "sculptured" and the edges of the septa are often very sinuous.

Subclass *Coleoidea*: squids, octopi.

(Class *Bivalvia*: clams, oysters, mussels, etc.)

(Class *Scaphopoda*: tusk-shells.)

Appendix 2 – The Geologic Time Scale

To facilitate reference to it, the geologic record is divided into eras and periods according to the time scale shown below:

Paleozoic era (600 - 225 million years ago)

Cambrian period	(600-500 million years ago)
Ordovician period	(500-440 million years ago)
Silurian period	(440-400 million years ago)
Devonian period	(400-350 million years ago)
Carboniferous period	(350-270 million years ago)
Permian period	(270-225 million years ago)

Mesozoic era (225 - 70 million years ago)

Triassic period	(225-180 million years ago)

Jurassic period (180-135 million years ago)
Cretaceous period (135-70 million years ago)
Cenozoic era (70 million years ago to the present)
Paleocene period (70-60 million years ago)
Eocene period (60-40 million years ago)
Oligocene period (40-25 million years ago)
Miocene period (25-11 million years ago)
Pliocene period (11-2 million years ago)
Pleistocene period (Two million years ago to the last ice age, or to the present, depending on the importance attached to Man)

Bibliographical Notes

A general account of mollusks and their evolution is provided in J. E. Morton, *Molluscs*, Hutchinson University Library (fourth edition, 1967). Supplementary background may be found in Moore, Lalicker, and Fischer, *Invertebrate Fossils,* McGraw-Hill, 1952.

Batten, Rollins, and Gould, "The Adaptive Significance of Gastropod Coiling," pp. 405-6 of *Evolution*, vol. 21, 1967, is an extremely informative note on the evolution of torsion.

CHAPTER 6

Biological Clocks

WHEN discussing adaptation, we speak of the best strategy among the available alternatives. What are the alternatives? A physicist or engineer might ask what is possible with the available energy or information: we have seen how such an approach clarifies the issue and sometimes answers our questions. An historian, on the other hand, might ask how the history of an evolving line has shaped the range of evolutionary alternatives presently open to it. These two attitudes are hardly exclusive: an historian can sometimes explain why one alternative costs as much as it does; why, for example, it might cost the cephalopod a nearly impossible amount of information to build a conispiral shell. Yet the temperaments of physicist and historian are so very different! The physicist is accustomed to knowing what he does and does not understand; he is usually capable, moreover, of phrasing his understanding in the language of mathematics. The historian, on the other hand, is hardly more than a poet. He knows neither the bounds of his understanding nor its precision: he depends on an intuition not yet sufficiently formed to be mathematized. To communicate his impressions of order he must *select* and *arrange* his subject matter so that his ideas will form in the minds of his readers. It is no accident that mathematics is absent from our historical chapters, this one and its predecessor.

How does the history of an evolving line shape the alternatives presently open to it? A single "accident of history" such as the development of torsion can be a turning point of evolution, determining the future roles of whole classes of organisms. But history is not all accident. Historical understanding can enable us to predict how organisms will solve certain environmental problems. One of the most striking examples of this is provided by the processes an organism uses to measure time, and the uses to which these processes are put.

There is abundant evidence that an organism, or some of its physiological processes, can measure time. A mimosa tree, which folds its leaves at night, will continue to do so when placed in a cave where darkness is constant and the temperature unchanging. If a bottle of developing fruit-fly pupae is taken from its place by the laboratory window and thrust into a dark constant-temperature room, there will be a daily rhythm of emergence from the pupal cases. If a cockroach or a mouse is placed under these conditions in a cage with a running wheel so arranged that every time the animal turns it round a mark is placed on a moving strip of paper, the marks will demonstrate a daily rhythm of activity. The rhythm is quite precise, and its period is remarkably independent of temperature: we thus speak of it as a *clock* telling the organism the proper time for its activities. The rhythm's period differs from twenty-four hours, suggesting that the clock expresses an internal rhythm of the animal rather than some influence of the earth's rotation: this clock must therefore be adjusted to the external rhythm of day and night. Such internal rhythms, called *circadian* in view of their *approximate* 24-hour period, have been found in all organisms studied save the bacteria and blue-green algae. Circadian rhythm is manifest in the rising of *Euglena* to the water surface, and in the susceptibility of rats to poisons; in changes of the chemical characteristics of enzymes, and in the variations of body temperature in hibernating bats. Circadian rhythm is indeed a basic expression of the physiology of living things.

How does the clock adjust to the cycle of day and night? The rhythm of light and dark is the surest way for the organism to judge this cycle: temperature, pressure, humidity, and the like are all subject to confusing and unpredictable fluctuations. And indeed, light is the only natural stimulus which can reset the clock*. The only other means known for altering the clock's expression are artificial chemical stimuli which interfere in wholesale fashion with protein synthesis, etc.

What does a light signal do? If one subjects fruit flies developing in the dark to a fifteen-minute light signal, the rhythmicity of the emergences will be disturbed for several days afterwards. Eventually the rhythm is re-established, but now the flies emerge a few hours earlier or a few hours later each day than they would have, had they not received the signal. It is as if the rhythm were advanced or delayed by the light signal, or, in other words, as if the light signal had "reset" the organism's clock. The degree of delay or advance depends on the stage of the rhythm at which the light signal arrives: we can tell, so to speak, what time the flies think it is by the effect of a fifteen-minute light signal on their emergence rhythm.

To tell what time flies in complete darkness think it is, we let the period between successive "emergence peaks" represent twenty-four "circadian hours" (which might be twenty-three earthly ones) and set the time of the emergence peak to be the time it would occur if dawn were at 6 a.m. and sunset at 6 p.m.

---

* In conditions of constant darkness, a rhythm can sometimes be reset by a temperature cycle.

We may then plot a "response curve," a graph which, for each circadian time, tells us how much a fifteen-minute light signal arriving at that time will advance or delay the rhythm of flies developing in otherwise complete darkness (Fig. 6-1). In general, signals received in the early "circadian morning" advance the rhythm, for under natural circumstances such early light would mean the organism's clock was slow: for similar reasons, signals received in the late circadian evening delay the rhythm.

The response curve suggests a simple model of the inter-

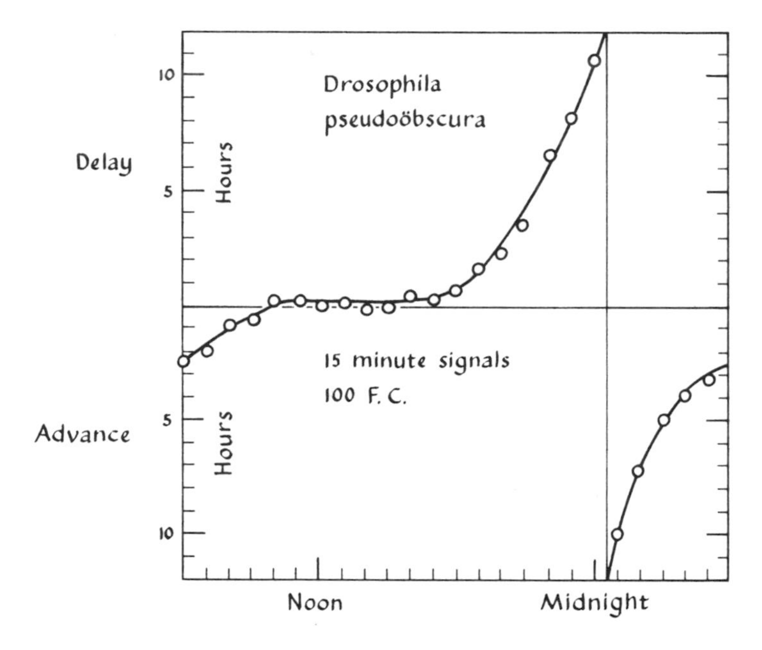

Fig. 6-1. Response curve of *Drosophila pseudoobscura* to fifteen-minute light signals. (After Fig. 8 of C. S. Pittendrigh and D. Minis, "The Entrainment of Circadian Oscillations by Light and their Role as Photoperiodic Clocks," p. 271 of the *American Naturalist*, vol. 98. Used with permission of C. S. Pittendrigh and the University of Chicago Press.)

action between a light signal and an otherwise "free-running" clock: we suppose a light signal immediately resets the clock the amount prescribed by the response curve. This model requires us to distinguish between a "pace-maker rhythm," which a light signal resets immediately, and the observed rhythm of behavior or emergence, which often requires several days to adjust to the resetting of the clock: the abundant success of the model confirms the validity of this distinction. If we seek to entrain the developing fruit flies to a periodic series of fifteen-minute light signals occurring nearly twenty-four hours apart, the signals must advance or delay the rhythm long enough each cycle to adjust it to their own cycle. We can predict the relation of the emergence peak to the light signal by asking what circadian time the signal must arrive each cycle to adjust the clock to the signal-rhythm. The model also allows us to predict the extent to which the clock's period can be entrained by (adjusted to) external rhythms: we find the clock can only be adjusted to periods near its own free-running period. Finally, the model tells us how the flies will read a pattern of two signals a day. The flies read these signals almost as if they were dawn and dusk: the model predicts, and observation confirms, that if the longer interval between signals exceeds thirteen hours, the insects always interpret it as night and the shorter interval as day-time.

What significance has the shape of a response curve? Nocturnal animals usually have free-running periods less than twenty-four hours, and the response curves show that signals in the circadian morning have relatively little effect, whereas signals in the circadian evening delay the rhythm the more the later they occur. The cycle is adjusted primarily by evening delays, so that activity is governed primarily by the time of sunset. *Drosophila pseudoobscura*, on the other hand, is active primarily at dawn and dusk. A signal near circadian sunrise advances the rhythm the more the earlier it occurs, whereas a signal near circadian sunset delays the rhythm the more the later it occurs. The phase of the rhythm as a whole is set by the balance between a.m. advances and p.m. delays, but morning activities will tend to stay near dawn, and evening ones near sunset.

Spectacular phenomena are associated with this internal clock. In former times, a good chronometer was vital to navigation, and the British Admiralty accordingly offered a great reward for the construction of a suitably accurate clock. Migrating birds navigate by the sun, and to know where the sun is they must know what time it is: birds, too, need a clock for navigation. A bird's navigational clock can be reset by an artificial light cycle: if birds are exposed to an artificial cycle for some while and then let out, they will orient themselves according to the time the artificial light cycle has led them to believe it is. If a starling with a 23½-hour free-running period is shut up in a dark cage for a week, upon release it will think the time of day is over three hours later than it actually is and orient accordingly. Thus the navigational clock is the basic circadian rhythm.

Not only are certain activities appropriate to certain times of day: certain activities are appropriate to certain times of year. Some flowers should bloom in the spring, and some in the fall. It pays summer's caterpillar to develop into a butterfly, while fall's caterpillar should wait through the approaching winter in a resting stage. Many organisms recognize seasons by the length of the day, knowing winter by its short days and summer by its long. The flower blooms in the daylength appropriate to its season, and the horticulturist can make it bloom when he wishes by subjecting it to artificial days of the proper length. Likewise, the length of day decides whether an insect develops or diapauses (enters a resting stage).

Do organisms use their circadian rhythms to measure the length of the day? Fruit flies emerge from their pupae before dawn on short days and after on long: stages of their rhythm which are in the dark on short days are illuminated during long. Perhaps an insect develops rather than diapauses if a certain "critical" stage of its circadian rhythm is illuminated. Suppose we subject insect larvae living in a short-day regime to a single additional one-hour light pulse each day, and ask what time, if any, the pulse stimulates development. If development depends on the illumination of a "critical" stage of the rhythm in the circadian morning, a light pulse will stimulate development if it is interpreted as sunset of a sufficiently long day to warrant it, or if it is interpreted as dawn of a day just long enough to warrant it. Notice that if the dawn signal falls earlier, it might still be interpreted as dawn of a long day, but the critical stage of the rhythm would then come after the signal, and thus be in the dark. These predictions hold for the one bollworm studied: this bollworm does use his clock to measure the length of the day.

Circadian rhythms thus *prejudice* their bearers' response to other problems of time measurement: the presence of clocks renders inevitable certain responses to these problems. Was the evolution of circadian rhythms inevitable in the first place? We are too ignorant of physiology to answer certainly, but plausible arguments do spring to mind. Biochemical processes cannot all go on at once: they must be ordered. Moreover, in small, transparent organisms there may be some reactions which should only be allowed to take place at night. The phasing of reactions to light and darkness may have been a more urgent business long ages ago, when the atmosphere screened out less of the sun's ultraviolet radiation. The organism perhaps had to develop a means by which the cycle of night and day would impress a twenty-four hour rhythm on its physiological processes. Selection would favor the development of an autonomous rhythm as opposed to a process requiring expenditure of energy to direct and control it. A similar principle is exemplified in the spherical shape so many micro-organisms possess, and the resemblance of other micro-organisms, and some larger animals, to forms of drop and splash (Fig. 6-2). These forms were not determined by the forces of surface tension which shape drop and splash; they were selected. In a world where surface tension is important,

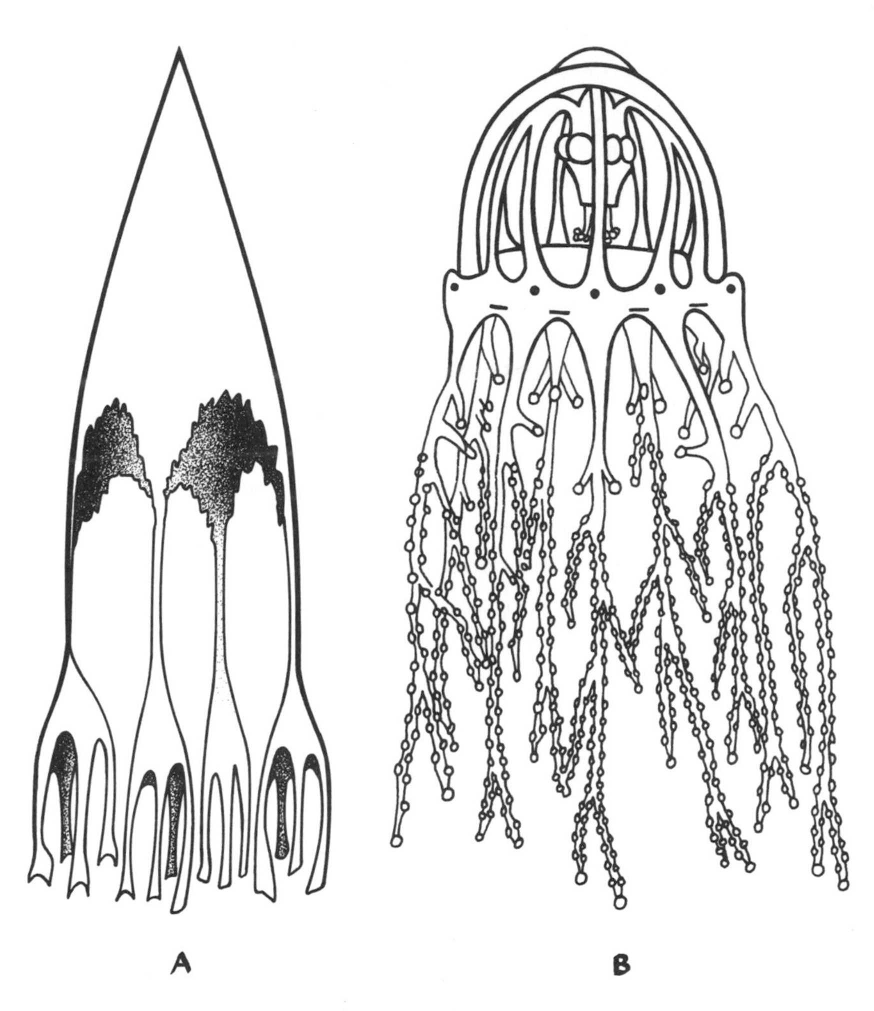

Fig. 6-2. A. A drop of fusel oil falling through paraffin.
B. The medusoid *Cordylophora*.
(After Figs. 120B and 121, no. 2, of D'Arcy Thompson, *Growth and Form*, vol. 1, 2nd edition, with permission of the Cambridge University Press, Cambridge, England.)

these forms were subject to the least deforming stress; they were the stablest and most "natural," requiring the least energy to build and maintain. Apparently it pays to "give in gracefully" to the forces of one's environment. In this connection it is perhaps worth noting that a flash of light at the proper circadian time has an unusually strong effect on the rhythm of *Drosophila pseudoobscura*. This fly lives in very dry areas: to avoid dessication, it must restrict its activity to dawn and dusk, even though this entails interference with its natural rhythm. Fruit flies (and other organisms) not requiring such a delicate adjustment of their activity grant their rhythms more autonomy, as if some advantage accrued to doing so.

Circadian rhythms may have been constructed from "feedback oscillations." A thermostat is the prototype feedback device: it turns the furnace off when temperature rises above the desired level and turns it on again when temperature falls below the optimum. The thermostat causes a slight temperature oscillation by not turning the furnace on until the house has cooled *detectably* below the optimum, and not turning it off again until the temperature is detectably too high. Biochemical feedback oscillations also occur, resulting from the imperfect adjustment of enzyme activity to the concentration of the products whose manufacture they catalyze. Because different feedback oscillations take place in the same chemical medium, they influence each other. Any of you who have marched in formation know how one can keep time to the rhythm of marching feet, and how, when each person adjusts his step to the sound of the whole, the formation moves rhythmically without anyone calling time. Feedback rhythms may also interact to form a composite rhythm more accurate than any of its components: perhaps organisms were favored in which a multitude of feedback rhythms interfered constructively to form a basic circadian rhythm. Computer simulations suggest that relatively slight adjustments in the interactions of the feedback oscillations could greatly change the period of the composite rhythm: the fact that circadian rhythms have much longer periods than some (most?) of their presumed biochemical components does not necessarily present difficulties for our story.

Composite rhythms may also have developed from feedback oscillations at a very different level of biological organization. Norbert Wiener discusses an electrical rhythm in the brain with a frequency of about ten cycles per second: like the circadian rhythm it orders processes in time. A neuron fires if and only if it receives simultaneous impulses from a certain number of its neighbors. It receives impulses only at a certain stage of the rhythm: it therefore "judges" whether impulses are simultaneous by whether they arrive during the same cycle of the rhythm. A more precise electrical analysis of the brain reveals a great many rhythms, many with frequencies near ten cycles per second, some with quite different frequencies, but surprisingly few with frequencies only moderately different from the peak. This spectrum of frequencies suggests that the major rhythm of ten cycles per second entrains rhythms with neighboring frequencies, or that rhythms near this frequency "pull together," entraining each other to produce the resultant. One may adjust the resultant rhythm to a nearby frequency by flickering light of that frequency into a person's eyes, just as one can entrain a circadian rhythm by an appropriate schedule of light signals. There is some hint that our resultant rhythm is caused by interactions between feedback oscillations. It seems nature approaches certain problems of temporal organization in a very uniform way indeed.

What has all this to say about the predictability of evolutionary process? In the preceding chapter, we showed that selection would favor conispiral coiling in crawling animals with permanent shells. If conispiral coiling were certain to occur among the ancestral mollusks, it was certain that eventually most crawling mollusks would be descended from conispiral stock. Surely, if predators and shelled forms of earthly dimensions have evolved on another planet, conispiral shell forms have evolved there too: indeed, a visiting shell collector should expect to find there most of the shell forms which so delight him here on earth. The evolution of circadian rhythms also appears to have been inevitable: surely this is indicated by the evolution of very similar rhythms to meet problems of temporal organization posed on a very different time scale in the human brain. In these examples we can only agree with Huxley that, having seen the result, we are convinced evolution could have taken no other path.

Analysis of the alternatives open to an evolving line sometimes enables us to predict not only the end results of evolution, but something of the branching patterns of the evolutionary tree leading to these results. Acquaintance with the symmetry of ancestral mollusks shows what a difficult business the development of torsion and conispiral coiling must have been. Yet these developments were central to the exploitation of molluscan possibilities. The first occurrence of torsion was therefore destined to be an "accident of history" determining the evolutionary future of entire classes of organisms: being such an unusual development, its first "successful" occurrence would be fully exploited before it could occur again. If we define the gastropods as that group of mollusks undergoing torsion during larval development, they are very probably monophyletic: that is to say, they are very probably descended from a single ancestral population. We would expect any group whose evolution hinged on the occurrence of such an accident to be likewise monophyletic.

A monophyletic pattern of development, however, could as easily result from an ecological bottleneck as a physiological one. It could well be that terrestrial vertebrates could only develop from ancestors living in seasonal ponds, because the stagnant conditions often encountered in such ponds were the only circumstances which would force aquatic animals to evolve lungs. From a fish's standpoint, such seasonal ponds offer a marginal way of life which can support only a few species. Thus, even though there may be no physiological stumbling block obstructing a fish's evolution into a land animal, land vertebrates as a group might be monophyletic because "there are so few possible ancestors to choose from."

On the other hand, there is no apparent obstruction to using one's circadian rhythm to tell the season of the year: one would hardly claim common ancestry for two metazoans which happened to tell seasons the same way. Taxonomists often assume their goal is to classify organisms into monophyletic groups: they wish their classification to be a family tree, indicative of degrees of common ancestry. In practice, however, they classify by structure: gastropods are defined as those mollusks undergoing torsion during larval development, rather than those mollusks descended from Cambrian pleurotomariids. Our previous discussion suggests that such a structurally defined group will be monophyletic only if its origin depended on an accident which was unlikely, but could have happened in a number of groups, or if there were only a few species with a form or way of life that would lead to evolution of the group. The taxonomists' insistence on monophyletic classification has accordingly led to difficulties which are perhaps best illustrated with the origin of mammals. As the class *Mammalia* is presently defined, at least four separate lines of reptiles, all belonging to the same order *Therapsida*, evolved mammals: this class is accordingly polyphyletic. Reed wants the *Therapsida* included among the mammals, arguing that warm-bloodedness evolved in the *Therapsida*, and that warm-bloodedness was what precipitated the increased activity and consequent skeletal developments now held to be characteristic of the mammals. However, there is no reason to suspect either a physiological or an ecological stumbling block to the evolution of warm-bloodedness: perhaps many Mesozoic reptiles other than therapsids were warm-blooded. In general, only the fact that so many taxa have evolved from "accidents of history" or have arisen from ancestors living a "marginal" way of life, and the fact that the origins of so many other taxa are obscure, permits the possibility of a strict phylogenetic classification (classification by ancestry) to be entertained.

Remember, however, that neither the results of evolution nor the details of the phylogenetic patterns leading to them are invariably predetermined. Although the convergence of marsupial and placental moles suggests a predictability in the evolution of mammalian burrowers, and the convergence of marsupial and placental wolves (Fig. 6-3) suggests the same for a certain grade of tetrapod carnivore, we must remember that though antelope and kangaroo lead ways of life quite as similar, they are very different. We must understand development and functional morphology better than we do before we will know what aspects of evolution are deterministic: that is to say, what aspects are predictable and amenable to general theory.

Problem

Consider a fruit fly with the response curve shown in Fig. 6-1.

a) Suppose we give it fifteen-minute light signals daily at 4 a.m. and 2 p.m. (our time), keeping it otherwise in constant darkness. If each light signal resets the animal's clock immediately, what time will the animal think it is at 4 a.m. and 2 p.m. when the clock reaches equilibrium? Assume the animal thinks it is 4 a.m. when the first signal is given.

b) Repeat the analysis, supposing now that signals are given at 4 a.m. and 8 p.m.

Solution: Start by making a table showing, for each hour of the animal's "subjective day," how a light signal will reset its clock.

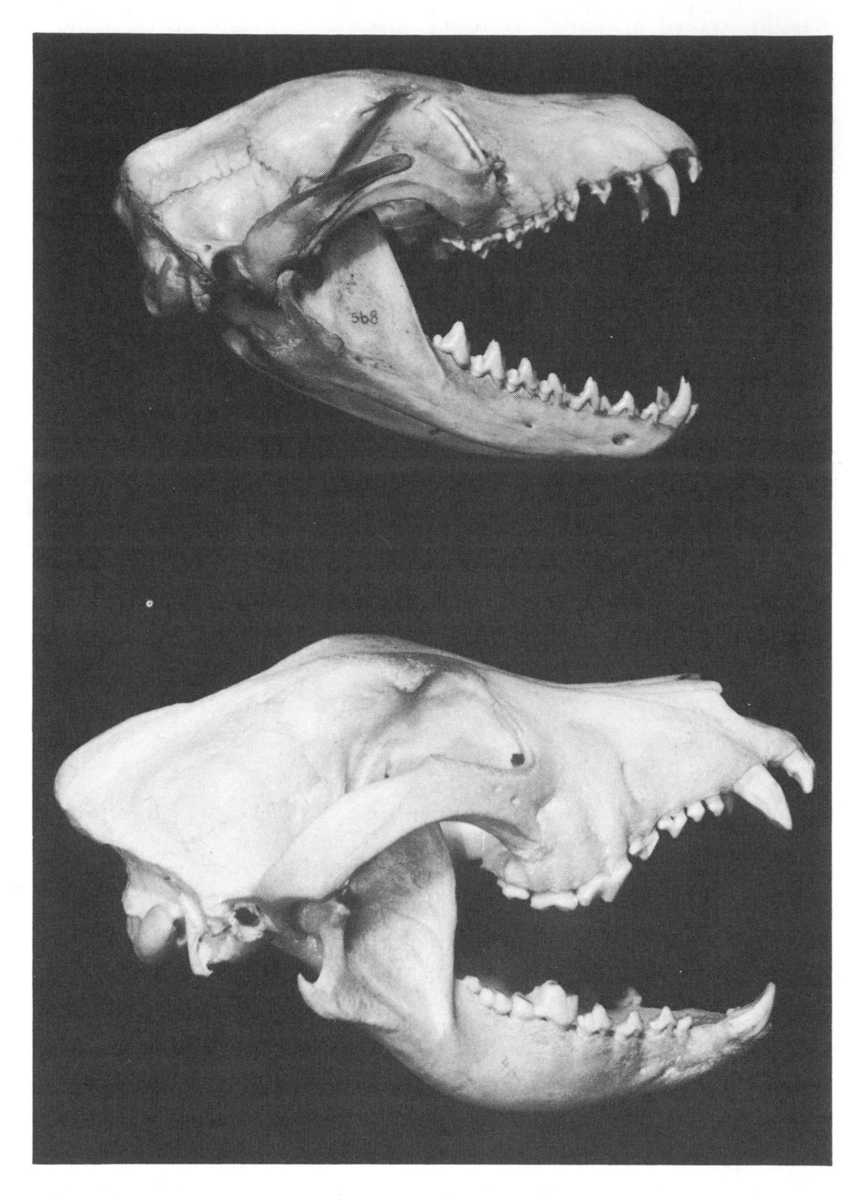

Fig. 6-3. Skulls of dog (bottom) and Tasmanian wolf (top). (Photograph by D. Baird.)

ANIMAL'S TIME

Before	*After*	*Before*	*After*	*Before*	*After*	*Before*	*After*
6 a.m.	8 a.m.	Noon	Noon	6 p.m.	5 p.m.	Midnight	1 p.m.
7 a.m.	9 a.m.	1 p.m.	1 p.m.	7 p.m.	5 p.m.	1 a.m.	11 a.m.
8 a.m.	9 a.m.	2 p.m.	2 p.m.	8 p.m.	5 p.m.	2 a.m.	9 a.m.
9 a.m.	9 a.m.	3 p.m.	3 p.m.	9 p.m.	5 p.m.	3 a.m.	8 a.m.
10 a.m.	10 a.m.	4 p.m.	4 p.m.	10 p.m.	3 p.m.	4 a.m.	8 a.m.
11 a.m.	11 a.m.	5 p.m.	5 p.m.	11 p.m.	2 p.m.	5 a.m.	8 a.m.

a) The first 4 a.m. signal shifts the animal's clock to 8 a.m. The first 2 p.m. signal arrives ten hours later, when the animal thinks it is 6 p.m. and sets his clock back to 5 p.m. The next 4 a.m. signal comes fourteen hours later, when the animal thinks it is 7 a.m. and shifts the clock to 9 a.m. The next 2 p.m. signal arrives at 7 p.m. "subjective time," shifting its clock back to 5 p.m., and causing a repetition of the cycle. At 4 a.m. the animal thinks it is 7 a.m.; at 2 p.m. the animal thinks it is 7 p.m.

b) When its clock reaches a steady state, the animal will think it is 8 a.m. when it receives the 8 p.m. signal, and 5 p.m. when it receives the 4 a.m. signal: it thinks the longer interval is night.

Bibliographical Notes

The role of historical analysis in biology is discussed in C. S. Pittendrigh, "On Temporal Organization in Living Systems," pp. 93-125 of the *Harvey Lectures*, vol. 56, 1961. This paper is also the best summary of the work on biological clocks. The use of clocks to tell the season is discussed in C. S. Pittendrigh and D. Minis, "The Entrainment of Circadian Oscillations by Light and their Role as Photoperiodic Clocks," pp. 261-294 of the *American Naturalist*, vol. 98, 1964, and C. S. Pittendrigh, "The Circadian Oscillation in *Drosophila pseudoobscura* pupae: a model for the Photoperiodic Clock," pp. 275-307 in *Zeitschrift für Pflanzenphysiologie*, vol. 54, 1966. The latter has a final section on the origin of biological clocks. Unfortunately, these two papers are far more difficult than the Harvey Lecture.

The notion that animal forms evolve to minimize environmental stress is discussed in G. E. Hutchinson's essay, "In Memoriam: D'Arcy Wentworth Thompson," reprinted in G. E. Hutchinson, *The Itinerant Ivory Tower*, Yale University Press, 1953.

Wiener discusses brain rhythms in the last chapter of *Cybernetics* (second edition, M.I.T. and Wiley, 1961). His analysis is not universally accepted.

A. T. Winfree discusses how oscillators may entrain each other to form a collective rhythm in a fascinating but quite mathematical paper, "Biological Rhythms and the Behavior of Populations of Coupled Oscillators," pp. 15-42 in vol. 16 of the *Journal of Theoretical Biology*, 1967. This model cannot yet "explain" biological clocks, but it brings a fresh approach to the subject.

C. A. Reed discusses the problems of classification in "Polyphyletic or Monophyletic Ancestry of Mammals, or, What is a Class?" pp. 314-322 of *Evolution*, vol. 14, 1960. This paper is perhaps the best attempt to explain an art which many feel is communicable only by example: most attempts to reduce taxonomy to a verbally communicable science deny any explicit connection between taxonomy and evolutionary history. For further discussion of the purpose and role of taxonomy, see the symposium edited by Haywood and McNeill, *Phenetic and Phylogenetic Classification*, published by the Systematics Association (London, 1964).

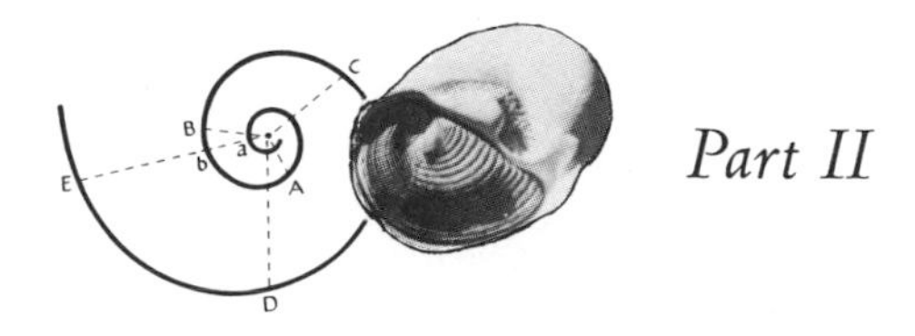

Part II

ECOLOGY

To penetrate the secret of adaptation, one must learn what gives the advantage to one species over another in a competition. Each role in life determines a different set of "rules of competition," and thus a different criterion of fitness: some species survive because they can crowd out competitors; others because they can starve them out. To learn how these roles are defined, we must learn how communities of living things are organized, how they function, and how they evolve.

CHAPTER 7

Competition

THE LAST section dealt with the alternatives open to an evolving line. How are these alternatives judged? What makes one animal more fit than another for a way of life? Answers to such questions come from ecology. This science is also concerned with the ways different plants and animals interact. Patterns of organization have evolved in these interactions, so that it is appropriate to treat the association of plants and animals living together in a region as a functional unit in its own right, as a *community*. These topics are all related, but it is convenient to distinguish a theory of competition from a theory of communities.

Competition theory hangs on the simple remark, almost too obvious for serious attention, that if two kinds of organism make their living in the same way, under the same conditions, one must be better at it than the other. This leads the biologist to the principle of competitive exclusion: if two species living in the same place make their living in the same way, one will replace the other. This principle explains a host of ecological phenomena. It tells us why we never find two species living the same way in the same community. The different warblers of a forest feed in different parts of the trees, and in different ways; the predatory snails on a reef feed on different foods, and in somewhat different places. In general, closely related species living in the same area have different preferences of food or habitat that keep them out of each other's way: the preferences may concern humidity, as in bromeliad-breeding mosquitoes, or vegetation, as in sympatric African rhinoceroses. Competitive exclusion goes far to explain why we can distinguish the entities we call species: each species of a community must have its distinctive way of making a living, at which it is clearly better than any other.

The evidence that differences between species are designed to prevent competition is provided by *character displacement.* On the islands of the Galápagos, a seed-eating finch eats smaller seeds, and has a smaller beak, on those islands where a larger seed-eater is present. In the presence of the larger finch, it adapts to a role sufficiently different so that each may lead a way of life in which it is clearly superior. Many other examples are known of this phenomenon.

What property renders a species more fit for its way of life than any of its competitors? Can we measure the fitness of different species for a way of life in such a manner that the more fit species wins in a competition experiment? To order species according to fitness, we need an additional assumption beyond the principle of competitive exclusion,viz., if A excludes B and B excludes C, then A excludes C.

To measure fitness, we must know the manner of population regulation a way of life imposes on its occupants. What are the different modes of population regulation? Our first distinction, that between equilibrium and non-equilibrium species, is based on the prospects of individual populations of the species in question. If a population can make a living in the same place generation after generation, that is to say, if it finds living conditions sufficiently stable and reliable to assure it a relatively permanent future where it is, then it belongs to an equilibrium species. Such is the case for the association of spruces and bog-moss of the far north, living in a changeless succession of brief summers and harsh winters: their ancestors have lived in the same region, under the same conditions, ever since the last glaciation. So also for the trees of the great tropical hylaea, the rain forest you know from *Green Mansions*, and for the mammals and birds which live there. So also for the forest world our forefathers destroyed in settling this country, and that world of redwood and Douglas fir which is being destroyed in our own time.

If, however, a population is subject to extinction before it can fully exploit its environment, if the conditions of life are such that a population's future in any one locale is transient and temporary, then the population belongs to a non-equilibrium or opportunistic species. Non-equilibrium species depend on an ability to disperse rapidly from one place to another to take advantage of any turn of conditions permitting them to settle in a new locale; in short, they depend on their ability to exploit new opportunities as they occur. The *Paramecium* living in a puddle that will soon dry out; the *Euglena* that will be replaced by a competitor as soon as its pond cools slightly; the weeds in an abandoned barnyard, soon to be choked by a tougher but slower-growing grass; the growth that springs up in the sunlight where a great forest tree has fallen, soon to be overshadowed by this tree's seedlings—all these opportunistic species see the world as a highly unstable place, to be exploited when opportunity arises with no thought for the future, which belongs to someone else.

How do we measure the fitness of an opportunistic species? Consider as example two species of *Paramecium* competing for a network of temporary puddles. The competitive ability of each species is related to its ability to multiply during the lifetime of an individual puddle, and to the proportion of one puddle's population that lives to colonize other puddles. Imagine, to be specific, a puddle colonized by N_1 individuals of the first species and N_2 of the second. How many individuals of each species will this puddle contribute towards the colonization of other puddles? Suppose that over the puddle's lifetime the populations of species 1 and 2 multiply by factors F_1 and F_2, respectively; suppose, moreover, that when our puddle dries out, fractions λ_1 and λ_2 of the two populations survive to colonize other puddles. Thus our puddle contributes $\lambda_1 F_1 N_1$ individuals of the first species and $\lambda_2 F_2 N_2$ of the second to the colonization of other puddles. If $\lambda_1 F_1$ exceeds $\lambda_2 F_2$, then $\lambda_1 F_1 N_1 / \lambda_2 F_2 N_2$ exceeds N_1 / N_2: the ratio of the numbers of species 1 to those of species 2, in the entire network, will be increased. If $\lambda_1 F_1$ exceeds $\lambda_2 F_2$ in most puddles, the first species will displace the second. $\lambda_1 F_1$ *measures* the fitness of species 1: it determines the competitive ability of that species (see Appendix 1). The values of $\lambda_1 F_1$ and $\lambda_2 F_2$ will depend on the temperature, chemistry, etc. of the network of puddles: one species may be superior to the other in warm places, and inferior in cool.

What is the biological meaning of all this? Fitness of our *Paramecium* for its way of life depends on its ability to multiply and disperse in its chosen habitat. Any opportunistic way of life places a premium on these abilities. A population can multiply faster either by having more children per parent or by having them sooner. Opportunistic species emphasize the latter strategy: their individuals are characteristically short-lived. Organisms can disperse in many ways: seeds may be small and light to ride the wind, or sticky and hooked like stick-tights and cocklebururs to ride passing animals. Pond micro-organisms may disperse by riding the feet of birds, or by being eaten in one place and defecated in another, etc.: they usually have a stage in their life cycle adapted to resist unusual stress.

We may divide equilibrium species into two further categories: species limited by the abundance of prey or predators and species limited by suitable space to grow. The logarithmic growth rate of a space-limited species decreases with increase in its numbers: there will be an equilibrium population size at which the species just maintains its numbers. The population's size may affect its growth rate through the space occupied, or it may do so in subtler ways, as when the shade of a forest inhibits the growth of saplings, or when a yeast's alcoholic waste products slow its growth.

The simplest relation between growth rate and population size for a space-limited species is given by the logistic equation

$$d \log N/dt = r - aN,$$

where N is population size, r is the logarithmic growth rate of the population when its numbers are very low, and a is a constant reflecting the influence of the population's size on its growth rate. We may set $r = aK$ in the above equation and write

$$d \log N/dt = a(K - N),$$

$$dN/dt = aN(K - N)$$

(remember that $d \log N/dt = (dN/dt)/N$). A small population will increase asymptotically toward K, and a large one will decrease to this level (Fig. 7-1). K is the population size this environment can support in equilibrium. A solution to this equation is the familiar S-shaped curve, rising slowly at first, but with increasing rapidity until N exceeds $K/2$, and then leveling off again. Examples of this general type of curve are provided by phenomena as diverse as the growth of a man's weight during his lifetime and the increasing speed of a car accelerating from a stoplight. All these curves are characterized by lack of oscillation in the approach to equilibrium: space-limited populations may oscillate, however, if there is some delay in the adjustment of growth rate to population size, the growth rate at time t being determined by the population size, s time units

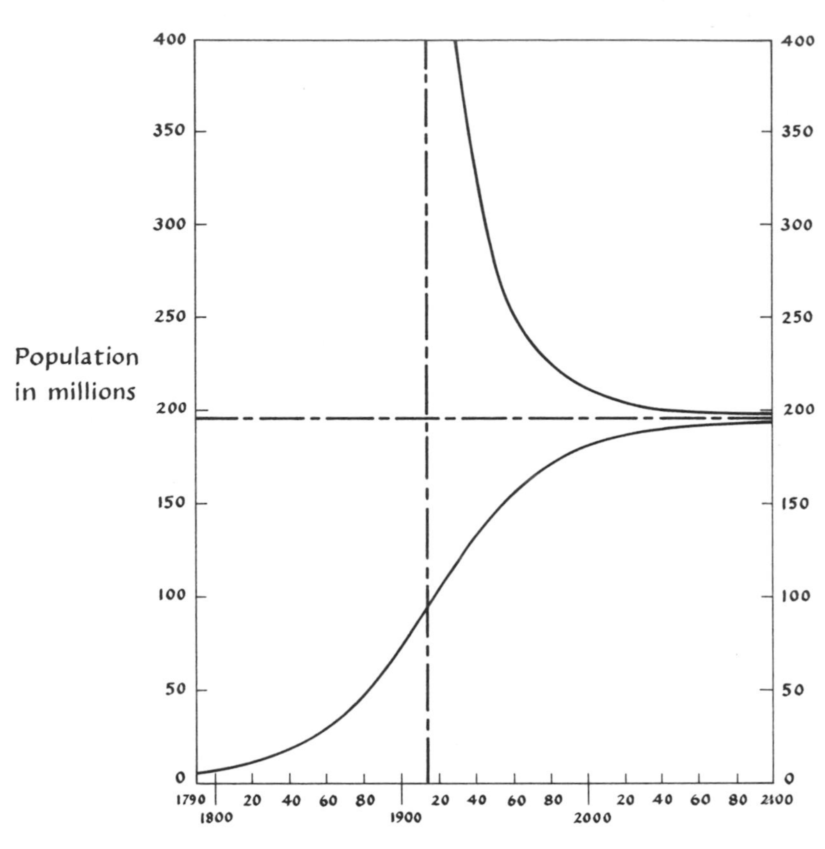

Fig. 7-1. The logistic law of population growth. The lower, S-shaped curve is the logistic curve fitted by Pearl and Reed to the U. S. census data from 1790 to 1910; the upper curve represents the hypothetical decline of an overcrowded population obeying the same equation. (From Fig. 4 of A. J. Lotka, *Elements of Mathematical Biology*, Dover Publications, Inc., New York, 1956. Reprinted through permission of the publisher.)

earlier. The logistic equation is distinguished by the linear relationship between $d \log N/dt$ and N: this relation is roughly obeyed by simple populations growing in culture-media, and quite precisely obeyed by the population of the United States between 1790 and 1940 (Fig. 7-2). Apparently the New Deal and the Second World War drastically altered the carrying capacity of this country: the U. S. population is now too high to fit its former curve. But it is worthy of reflection that the population of the U. S. over the first 150 years of independence fitted this equation better than most laboratory populations. It is as if a certain degree of complexity were required of ecological systems for them to exhibit orderly behavior.

How do we measure the fitness of space-limited species? Consider the competition of two yeasts for the same space. This is of some historical interest: the experimental studies of Gause on this subject provided the first "proof" of the principle of competitive exclusion. Suppose the two species of yeast are growing in an aerated chamber that holds a fixed amount of fluid, into which nutrient solution is added at a constant rate, and from which fluid (but no yeast) is withdrawn at the same rate, to balance the volume of fluid in the growth-chamber. The growth rates of these yeast populations depend on the alcohol concentration in their medium: they decline as it increases. The yeasts, however, make alcohol as waste products of their metabolism. The species which can tolerate the higher alcohol concentration is the only one which will persist, for its population will rise until the alcohol produced by the two species causes its competitor's extinction. Fitness for this experimentally created niche is thus measured by alcohol tolerance.

How may we abstract the essentials of this competition? Consider two space-limited species, and measure their populations in terms of the characteristics by which they crowd their competitors. The yeast populations of the preceding paragraph are best measured by their contributions to the alcoholic content of their medium. Similarly, we measure the populations of two barnacles competing for room to grow by the space they occupy, and the populations of two trees competing for light by

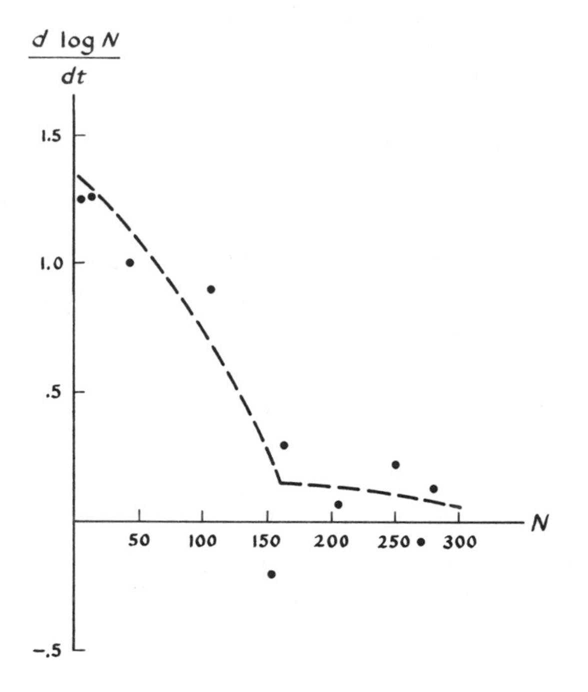

d log N / dt
1.5
1.0
.5
−.5
50
100
150
200
250
300
N

A

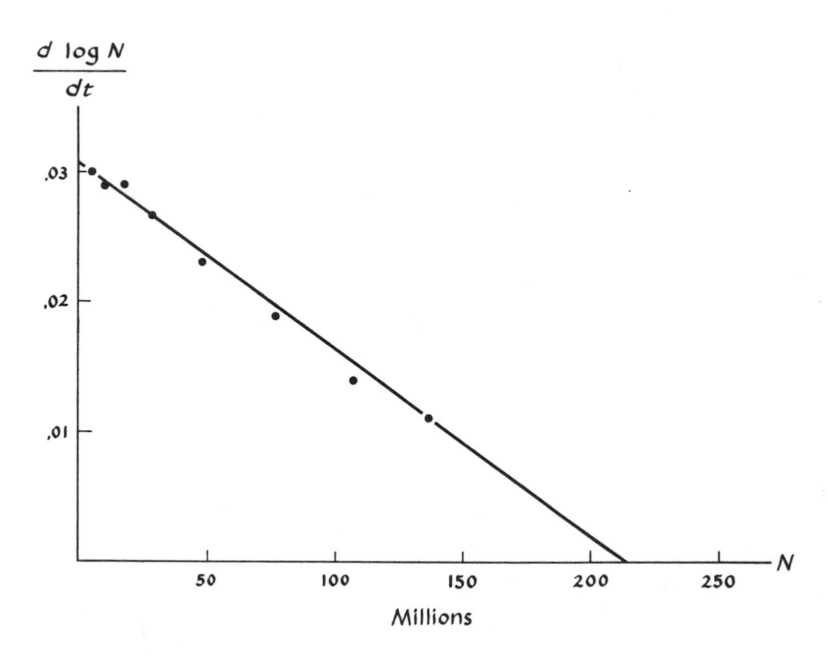

d log N / dt
.03
.02
.01
50
100
150
200
250
N
Millions
B

the shade they cast (which we estimate by the total leaf area they support). If we let N_1 be the population of the first species and N_2 that of the second, then $N_1 + N_2$ measures the "crowdedness" of the environment. If the logarithmic growth rate of each species is diminished in proportion to the crowdedness of the environment, then

$$d \log N_1/dt = a_1 [K_1 - (N_1 + N_2)],$$

$$d \log N_2/dt = a_2 [K_2 - (N_1 + N_2)].$$

K_1 is the maximum crowdedness species 1 can tolerate, and K_2 has the same significance for species 2; a_1 and a_2 are constants measuring the rates at which the populations respond to excess or deficit in crowding. To predict which species will win a competition, we subtract x times the second equation from the first, where $x = a_1/a_2$: we find

$$\frac{d}{dt} \log N_1 - x \frac{d}{dt} \log N_2 = \frac{d}{dt} \log (N_1/N_2{}^x) = a_1 (K_1 - K_2).$$

If $K_1 > K_2$, the ratio of N_1 to $N_2{}^x$, and thus the ratio of N_1 to N_2, will increase without limit, and species 2 will go extinct. If, on the other hand, K_2 exceeds K_1, species 1 dies out. The species with the greatest tolerance of crowding alone survives. The fitness of a space-limited species is measured by its ability to withstand crowding in its chosen habitat: its capacity to survive is measured by its ability to *crowd out* its competitors.

Now let us turn to those equilibrium species limited by "foodweb relationships," the availability of prey or the abundance of predators. Consider as example a system consisting of a single prey species and its predator: how are these populations

---

Fig. 7-2. $d \log N/dt$ graphed as a function of N for a laboratory culture of *Paramecium* (A) and for the population of the United States (B). The curve in (A) and the line in (B) indicate the trend of the data.

regulated? In the absence of predators the prey population increases, but the prey growth rate declines as the predators increase. The predators decrease in the absence of prey, but the predators' growth rate increases as the prey increase. Suppose the system starts with few predators and many prey. The prey increase, as there are not sufficient predators to limit them: the predators increase because of the abundance of prey. Eventually the predators become so numerous that the prey population decreases, but the predators will not themselves decrease until there are too few prey to support them. When this happens the predators decline until the prey increase sufficiently in abundance to support them once again, at which point the system has returned to its initial composition of few predators and many prey. This is a cyclical process in which each population oscillates, contrasting strikingly with the steady approach of space-limited populations to equilibrium.

How may we describe this process mathematically? It is simplest to assume that the logarithmic growth rate of the prey is diminished in proportion to the abundance of the predators, while the predator growth rate increases in proportion to the prey abundance (Fig. 7-3). Letting N_1 and N_2 be the abundance of prey and predator respectively, we may write (see Appendix 2)

$$d \log N_1/dt = r_1 - a_{12}N_2;$$

$$d \log N_2/dt = -r_2 + a_{21}N_1,$$

where r_1 is the logarithmic growth rate of the prey population in the absence of predators, $-r_2$ the predator growth rate in the absence of prey, a_{12} a constant expressing the effect of predator abundance on prey growth rate, and a_{21} a constant expressing the effect of prey abundance on predator growth rate. If we set r_1 equal to $a_{12}Q_2$ and r_2 equal to $a_{21}Q_1$, our equations become

$$d \log N_1/dt = a_{12}(Q_2 - N_2);$$

$$d \log N_2/dt = a_{21}(N_1 - Q_1).$$

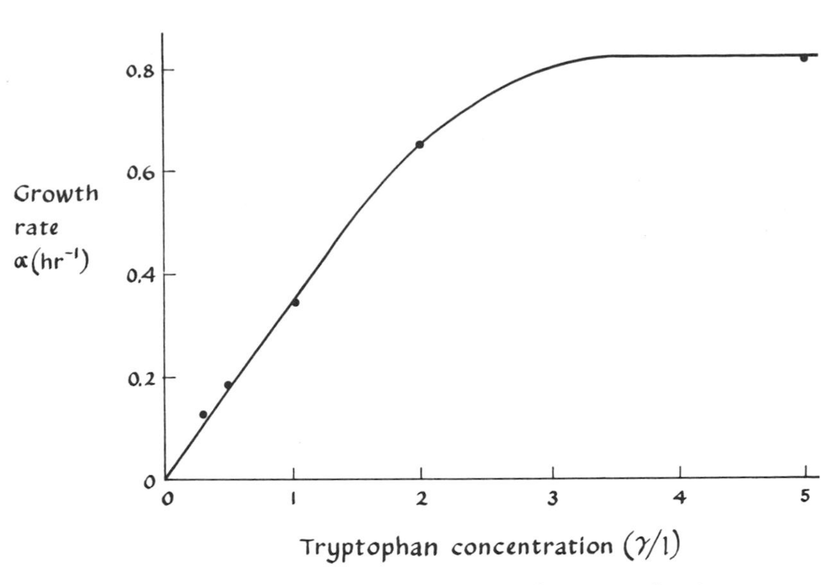

Fig. 7-3. The logarithmic growth rate (per hour) of a tryptophan-requiring strain of *Escherichia coli* as a function of tryptophan concentration. (From Fig. 1 of A. Novick, "Some Chemical Bases for Evolution in Micro-organisms," in A. A. Buzzati-Traverso, *Perspectives in Marine Biology*, University of California Press, Berkeley, 1958. Reprinted by permission of the Regents of the University of California.)

The predators thus increase if the prey exceed the threshold abundance Q_1 required for predator maintenance, and decrease otherwise; similarly, the prey increase only if the predator abundance does not exceed Q_1. Notice that

$$a_{12}(Q_2 - N_2)(d \log N_2/dt) + a_{21}\,(Q_1 - N_1)(d \log N_1/dt)$$

$$= 0;$$

$$a_{12}(Q_2 \log N_2 - N_2) + a_{21}\,(Q_1 \log N_1 - N_1) = \text{const.}$$

If we graph the second equation on a plane with axes N_1 and N_2, we obtain a closed curve, the topological equivalent of a circle (Fig. 7-4). The point (N_1, N_2) representing the population composition of our system is restricted to this curve, and must move round it in one direction, executing the same cycle over and over again.

A more refined theory would incorporate the fact that, even if there were no predators, the prey population could not

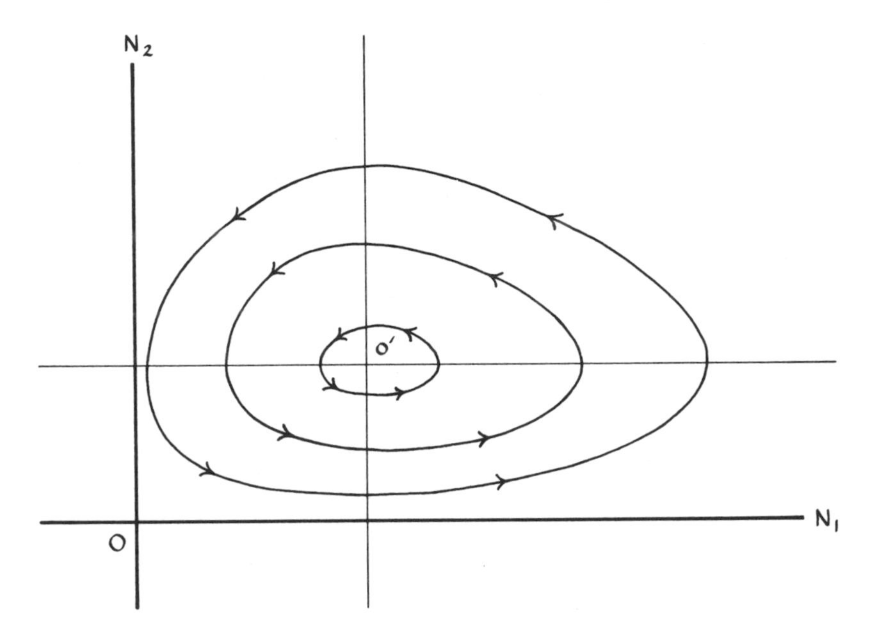

Fig. 7-4. The relation between the number of prey (N_1) and the number of predator (N_2) in predator-prey oscillations of different amplitude. O' is the point Q_1, Q_2 representing the equilibrium number of predator and prey: the smaller the oval about O', the lower the amplitude of the oscillation it represents. The lines marked x and y are the "isoclines" $dN_1/dt = O$ $(N_2 = Q_2)$ and $dN_2/dt = O$ $(N_1 = Q_1)$, respectively. (From Fig. 13 of A. J. Lotka, *Elements of Mathematical Biology*, Dover Publications, Inc., New York, 1956. Reprinted through permission of the publisher.)

increase indefinitely: the prey growth rate would eventually be depressed by the effects of crowding. These "crowding effects" damp out the oscillation. In a fluctuating environment, however, such populations oscillate at an amplitude set by the balance between environmental disturbance and the damping effects of crowding.

What is the evidence for such predator-prey oscillations? Many have attempted to produce a predator-prey oscillation in the laboratory. Usually, the predators eat the prey to extinction; sometimes, however, the predators starve to death before eating all the prey. If, however, the experimental environment is so complex that the predators can never find all the prey, while there are always some predators finding enough to eat, oscillations will occur. In Alaska, a natural oscillation has been described involving lemmings and the grass they eat; the oscillation has a period of three or four years. Here again, a complex natural association exhibits more orderly behavior than simple laboratory systems.

How do we measure the fitness of a food-limited species? Consider a species of bacterium growing in a chemostat. This is a chamber, into which nutrient solution is added at a constant rate, and from which fluid, with the bacteria it contains, is drawn off at an equal rate to balance the volume in the chamber. The higher the bacterial population, the lower the concentration of nutrient in its surroundings. The population grows until the nutrient level is so low that the population can only replace the bacteria drawn off with the fluid: the population is thus food-limited. If a second species is introduced to compete with the first, the species requiring the lower nutrient level for its maintenance will multiply until it starves out its competitor. And, in general, food-limited species depend on their ability to starve out their competition.

How may we "prove" this? Consider two species competing for a single prey. Let y be the abundance of prey, and N_1

and N_2 the population sizes of the two predator species. If the logarithmic growth rate of each predator increases in proportion to the prey's abundance, then

$$d \log N_1/dt = a_1(y - R_1);$$

$$d \log N_2/dt = a_2(y - R_2).$$

R_1 is the abundance of prey the first predator requires for subsistence, and R_2 is that required by the second; a_1 and a_2 are constants measuring the rates at which the predator populations adjust to prey abundances (a_1 and a_2 play the role that a_{21} plays in the equations for the predator-prey oscillation). Subtracting x times the second equation from the first, where $x = a_1/a_2$, we obtain

$$d \log N_1/dt - x(d \log N_2/dt) = \frac{d}{dt} \log (N_1/N_2{}^x)$$

$$= a_1(R_2 - R_1).$$

The ratio of N_1 to $N_2{}^x$ increases indefinitely if R_2 exceeds R_1 and decreases otherwise, implying that the successful species is the one requiring the lower abundance of prey for its maintenance.

We have discussed three extreme cases, the strictly opportunistic, the strictly space-limited, and the strictly food-limited species. We could have discussed the strictly predator-limited species, which evolves to maximize the predator population it can support, so that its predators will eat all its competitors to extinction. There is, of course, a full array of intermediates between these extremes. Opportunistic species come in all degrees, from those which exploit fleeting opportunities to those which strike a delicate balance between multiplying quickly and exploiting their opportunity for some little time afterwards (see Appendix 3). The different degrees of opportunism are evident in the progress of a plant succession, each stage of which is slower-multiplying and longer-lasting than the one before. There is a similar array of intermediates between food-limited and space-limited species; moreover, the dynamics of the two are not always as different as we have suggested (see Appendix 4). However, each way of life, or as the ecologist would say, each *niche*, is characterized by a combination of limiting factors, its degree of permanence, its pattern of predator pressures, its food and space limitations. Surely it is possible to rank species according to their fitness for some given niche, according to their ability to exploit its conditions.

Can we generalize about the factors limiting different kinds of organisms? A paper by Hairston, Smith, and Slobodkin may help to sort things out. This paper is exceedingly unpopular among ecologists, for no apparent reason other than its simplicity and generality. These authors think plants cannot be herbivore-limited, since so many leaves fall uneaten to the ground: plants must be limited by sunlight and water. If plants are not herbivore-limited, then the animals which eat them cannot be food-limited. What limits herbivores? The story of the Kaibab deer suggests an answer. The Grand Canyon abuts on a great mesa, about 8,000 feet high, which is capped by a quite beautiful forest. This forest harbors the Kaibab deer, and once harbored a host of pumas and wolves as well, until people thought to increase the numbers of deer by killing off their predators. So they killed all the pumas and wolves, and what was a herd of 5,000 deer increased to 100,000, whereupon they nearly all starved to death in two successive winters, seriously depleting the forest in the process. It appears that before the Europeans came along, there were a fifth as many deer as the forest would support. Apparently the deer were predator-limited in those days; their predators must accordingly have been food-limited. Hairston, Smith, and Slobodkin think most herbivores are predator-limited and most carnivores are food-limited. Limitation by sunlight and water is a form of space limitation: should we accordingly think plants space-limited and animals limited by foodweb relationships?

This rule is not universally valid. Grasslands may exist either because grasses like the prevailing climate or because zebras eat tree seedlings. Those who come from the valleys around San Francisco know how effective cattle can be at converting open woods to grassland. The landscape there is one of rolling grassland dotted by oaks, a play of dark green against a lighter one in the spring-time, of dark green against gold in the summer and fall. Look again at the oaks, and you will find no young ones, only oaks old enough to have been there before the cattle came, for the cattle eat all the seedlings, as well as all the leaves, they can reach. As the old, spreading oaks are blown down one by one, the landscape becomes more and more a grassland. Animals sometimes limit plants in more obvious ways: we have already mentioned a vegetation limited by the lemmings feeding upon it. The distinction between space-limited plants and foodweb-limited animals may also break down in subtler ways. Birds evolved territoriality to prevent overuse of their food supply: from an evolutionary standpoint birds are food-limited, but their dynamics are those of space-limited populations. Despite these objections, I feel the rule that plants are space-limited and animals limited by foodweb relationships is a useful approximation, a helpful guide in our present ignorance.

What of Hairston's second distinction? Are carnivores food-limited and herbivores predator-limited? We don't know. Carnivores often have their own predators: lizards and birds eat spiders, and snakes eat lizards. What limits the lizards and spiders? Moreover, the richness of a forest's foliage does not imply that leaf-eating insects are predator-limited. A tropical forest has hundreds of kinds of trees. Each kind makes a characteristic poison in its leaves: a leaf-eating insect can only cope with a few of these poisons. The trees may flourish while insects are starving in the search for leaves they can eat.

There are a variety of ways to decide what limits an animal. We can experiment: what happens if its predators are removed? This is sometimes more practical than it sounds: one can build a screen over a patch of mussels and barnacles to keep out the snails that normally prey on them. One can observe: if each member of a species stakes out a private territory to feed in, as is the case with many birds and mammals, the species is probably food-limited. A predator-prey oscillation is prima facie evidence of a food-limited predator and a predator-limited prey. The techniques used to decide what limits an animal are sometimes quite clever. Snails and hermit crabs cannot be food-limited, since they co-exist even though hermit crabs eat nothing not eaten by some snail or other. R. Vance informs me that the hermit crabs of the Washington coast compete vigorously for shells, taking up nearly every one available and fighting for them when need arises. His experiments show that crabs readily eat hermits with ill-fitting shells. Thus snails and hermits are apparently both predator-limited. If one can show that the diversity of hermit crabs in a given area is governed by the diversity of shells available, the case will be even stronger.

Appendix 1

The theory of opportunistic species presented here represents current ecological thinking on the subject quite faithfully, but it is logically inconsistent. We assumed the N *Paramecia* colonizing a puddle would give rise to λFN individuals colonizing other puddles; these in turn would give rise to a total of $\lambda^2 F^2 N$ colonists, etc. Since the *Paramecia* cannot increase indefinitely, λF must decline eventually to 1: selection cannot "maximize" it. The fact that, if one species replaces another, the successful species has a higher value of λF *during the replacement* does not imply that λF measures species fitness: λF would be 1 for either species, were it alone.

A correct theory would transform the distinction between

opportunistic and equilibrium species into one of degree rather than kind. Such a theory might predict the number of stages in a plant succession and account for the degree of opportunism in each. Meanwhile we must settle for an inconsistent theory which, whatever its faults, does explain why an opportunistic way of life places a premium on the ability to disperse and multiply.

Appendix 2

We wrote the equation for the changes in prey abundance as

$$d \log N_1 / dt = r_1 - a_{12} N_2.$$

If we set

$$d \log N_1 / dt = \frac{1}{N_1} \frac{dN_2}{dt},$$

we may rewrite our equation as

$$dN_1 / dt = r_1 N_1 - a_{12} N_1 N_2.$$

If the logarithmic growth rate of the prey population declines in proportion to the abundance of predators, the prey individuals are eaten at a rate proportional to the product of the population sizes of prey and predator, which we may in turn assume proportional to the number of meetings between members of the two species. These equations thus represent the simplest possible predator-prey interaction: prey and predators meet at random, and a fixed proportion of these encounters leads to capture of a prey.

Appendix 3

The chemostat may permit an experimental study of different degrees of opportunism, and allow us to evaluate the extent to which the ability to multiply interferes with exploitation of an equilibrium environment.

The equation of bacterial growth in a chemostat may be written as

$$d \log N/dt = aR - r,$$

where N is the bacterial population size, R the nutrient concentration in the growth chamber, a a constant expressing the effect of nutrient concentration on bacterial growth rate, and r the bacterial death rate, which we assume is due entirely to the removal of excess fluid, and its contents, from the chemostat. Setting $r = aR'$, we obtain

$$d \log N/dt = a(R - R'),$$

where $R' = r/a$ is the nutrient concentration the bacteria require for maintenance. When $R = R'$, the rate r of nutrient replacement is exactly equal to the bacterial growth rate. Perhaps one can conduct replicate "evolutionary experiments" in chemostats which are identical except for the rate of nutrient renewal, to see how the equilibrium concentration R' of free nutrient in the chamber depends on the bacterial growth rate r. The study of bacterial evolution is a tricky business: the experiment is quite ruined if the bacteria "learn" to lower their death rate by growing as a slime on the chamber walls; and the nutrient requirements of bacteria in a chemostat often change by occasional large jumps, rather than by many small ones. But careful experimentation should tell us more about the difference between opportunistic and equilibrium species.

Appendix 4

Although the bacteria in a chemostat are food-limited, they behave in many ways as a space-limited population. Let the nutrient concentration be R and the bacterial population N. If a proportion r of the bacterial suspension is removed per unit time and replaced by fresh nutrient solution, then

$$dN/dt = aNR - rN,$$

where a expresses the effect of nutrient supply on bacterial growth rate. R obeys the equation

$$dR/dt = r(C - R) - bNR,$$

where C is the concentration of nutrient in fresh solution, and bR is the rate at which a single bacterium consumes nutrient.

A fixed nutrient inflow, such as the chemostat gives, stabilizes the bacterial population. The population may oscillate, but if it does, the oscillation damps out. Moreover, we may measure fitness either by the equilibrium concentration of free nutrient in the growth-chamber, or by the population's rate $r(C - R)$ of nutrient uptake: as the former decreases, the latter inevitably increases. Indeed, if we measure the bacterial population by its rate of nutrient uptake (these being the units in which it affects competitors), we find it behaves very like a space-limited population.

How stable is a bacterial population in a chemostat? The chemostat is in equilibrium if neither the bacterial population nor the nutrient concentration is changing, that is to say, if $dN/dt = dR/dt = 0$. This is so if

$$R = r/a;$$

$$N = (a/b)(C - r/a).$$

The equilibrium nutrient concentration in the chemostat is independent of the nutrient concentration C in fresh solution: if C increases, the bacterial population grows until it reduces the nutrient level in the chamber to r/a. What happens if N and R are slightly displaced from equilibrium? Let

$$R = r/a + x;$$

$$N = (a/b)(C - r/a) + y.$$

Then we may write

$$dN/dt = dy/dt = [(a/b)(C - r/a) + y](r + ax - r).$$

If y is far smaller than the equilibrium value of N, then

$$dy/dt = (a^2 x/b)(C - r/a).$$

Similarly,

$$dx/dt = - aCx - bry/a.$$

Differentiating the equation for dx/dt and substituting for dy/dt in terms of x, we obtain

$$d^2x/dt^2 + aC(dx/dt) + arx(C - r/a) = 0.$$

To see whether oscillations occur, let $x = e^{wt}$, where e is the base of the Napierian logarithms. Substituting this expression for x in the above equation, and remembering that $d(e^{wt})/dt = we^{wt}$, $d^2(e^{wt})/dt^2 = w^2e^{wt}$, we find that w must satisfy the quadratic

$$w^2 + aCw + ar(C - r/a) = 0.$$

The roots of this equation are real and negative: this model predicts a uniform approach to equilibrium.

Problems

1. (Review of logarithms.) Consider a group of biology graduate students. Knowing the consequences of differential reproduction, but forgetting the need for population control, they marry early. Suppose they multiply so quickly that the volume of people doubles every twenty years. Starting with a sphere a cubic yard in volume, how long before this mass of people engulfs the sun (in other words, how long before the sphere's radius exceeds 90,000,000 miles)? How long before the edge of this spherical mass of people is expanding outward at the speed of light? Assume a light-year is 1.0×10^{16} yards: ignore the relativistic contraction in your calculations.

Solution: If the volume of people doubles every generation (one generation being 20 years), then the volume at generation t is 2^t times that at generation 0. Thus we must find t such that

$$V_R = 2^t V_0$$

where V_R is the volume of a sphere of radius R, where R is $1760 \times 9 \times 10^7$ yards, and V_0 is one cubic yard. Since V_R is $(4\pi/3)R^3$, we have

$$(4\pi/3)R^3 = 2^t,$$

$$\log(4\pi/3) + 3 \log R = t \log 2,$$

$$t = [\log(4\pi/3) + 3 \log R] / \log 2.$$

Substituting the numbers, we find that the sun will be engulfed in 114 generations, or 2280 years.

We may use two methods to find how long it will be before our sphere is expanding at the speed of light. The first says this happens if the sphere's radius increases 20 light-years in a generation. Since the sphere's volume is $4\pi/3$ times the cube of its radius, the radius may be expressed as $\sqrt[3]{3/4\pi}$ times the cube root of its volume. The radius is expanding outward at the speed of light if

$$\sqrt[3]{3V_0 2^t/4\pi} - \sqrt[3]{3V_0 2^{t-1}/4\pi}$$
$$= 20 \times 10^{16} \text{ yards.}$$

Setting $\sqrt[3]{3V_0 2^t/4\pi}$ equal to $\sqrt[3]{2}\sqrt[3]{3V_0 2^{t-1}/4\pi}$, we may rewrite the above equation as

$$(\sqrt[3]{2} - 1)\sqrt[3]{3V_0 2^{t-1}/4\pi}$$
$$= 2 \times 10^{17}.$$

Thus

$$\log(\sqrt[3]{2} - 1) + (1/3)[\log(3/4\pi) + (t - 1)\log 2] = 17.3;$$
$$t - 1 = [51.9 + \log(4\pi/3) - 3\log(\sqrt[3]{2} - 1)]/\log 2.$$

The sphere's outward edge expands at the speed of light after 182 generations, or 3640 years.

We may assume, on the other hand, that the sphere's volume expands continuously, so that $V_t = e^{kt}V_0$ where e is the base of the "natural logarithms" and k is chosen so that $e^k = 2$. k is $\log_e 2$, or 0.693. At generation t, the sphere's radius is $\sqrt[3]{3V_t/4\pi}$. When docs

$$\frac{d}{dt}\sqrt[3]{3V_t/4\pi} = 2 \times 10^{17}?$$

Substituting $V_t = e^{kt}V_0$ in the above equation, we obtain

$$2 \times 10^{17} = \frac{d}{dt}\sqrt[3]{3e^{kt}V_0/4\pi} = \sqrt[3]{3V_0/4\pi}\frac{d}{dt}e^{kt/3}$$
$$= \frac{k}{3}\sqrt[3]{3/4\pi}\, e^{kt/3}.$$

Taking natural logarithms of both sides, we obtain

$$\log_e (k/3) + \frac{1}{3}\log_e (3/4\pi) + kt/3 = \log_e 2 + 17 \log_e 10.$$

We find once again that the radius of the sphere is expanding at the speed of light after 182 generations.

2. Find logistic curves fitting the population of the United States from 1790 to 1950 (Table One: U. S. Census data), and the number of *Paramecium aurelia* in a laboratory culture (Table Two: data from Gause).

Table One

year	N (no. of individuals)	$\log_e N$
1790	3,930,000	15.1842
1810	7,240,000	15.7951
1830	12,870,000	16.3704
1850	23,190,000	16.9592
1870	39,820,000	17.4999
1890	62,950,000	17.9579
1910	91,970,000	18.3370
1930	122,775,000	18.6258
1950	150,700,000	18.8308
1970	208,000,000	19.1530

Table Two

day	N	$\log_e N$
1	2	.6932
2	7	1.9459
3	25	3.2189
4	68	4.2195
5	168	5.1240
6	138	4.9273
7	190	5.2470
10	222	5.4027
11	280	5.6348
12	260	5.5607
13	300	5.7038

Solution: We must determine whether $\frac{d}{dt}\log_e N = r - aN$. For the human data, graph $\frac{1}{20}[\log_e N(t + 10) - \log_e N(t - 10)]$ against $N(t)$, where $N(t)$ is the geometric mean of $N(t-10)$ and $N(t + 10)$: that is to say, $\log N(t) = \frac{1}{2}[\log_e N(t + 10) + \log_e N(t - 10)]$. In looking up the logarithms, it is often most convenient to look up logarithms to the base 10 and multiply by 2.3026. The data required for the graph are given in Table Three; they are graphed in Fig. 7-2. These data are fitted tolerably well by the equation

$$d \log_e N/dt = .0311 - (1.545 \times 10^{-10})N.$$

For the *Paramecia*, graph $\log_e N(t) - \log_e N(t-1)$ against the geometric mean of $N(t)$ and $N(t-1)$, as in Fig. 7-2: the necessary data are given in Table Four.

Table Three

year	N	$d \log N/dt$
1800	5,334,000	.0305
1820	9,654,000	.0288
1840	17,280,000	.0294
1860	30,390,000	.0270
1880	50,060,000	.0229
1900	76,090,000	.0190
1920	106,630,000	.0144
1940	136,000,000	.0103
1960	177,000,000	.0161

Table Four

day	N	$d \log N/dt$
1.5	3.74	1.25
2.5	13.4	1.27
3.5	41.2	1.00
4.5	107	.904
5.5	152	–.197
6.5	162	.320
8.5	205	.052
10.5	249	.232
11.5	270	–.074
12.5	280	.143

3. The numbers of foxes caught by the Moravian Mission each year from 1834 to 1879 were, respectively, 390, 90, 150, 200, 400, 70, 700, 1500, 300, 600, 35, 100, 700, 50, 100, 150, 700, 25, 100, 400, 650, 120, 60, 100, 350, 20, 140, 140, 500, 250, 70, 75, 1200, 100, 60, 150, 500, 320, 80, 400, 575, 425, 180, 80, 400, 1100 (data compiled by Elton; these numbers read off from a graph in Allee, Emerson, Park, Park, and Schmidt). Treat these numbers as censuses of the fox population.

a) Does the population appear to be space-limited (i.e., is it regulated by its own numbers)?

b) Is there any time lag in this regulation?

Solutions: The population is density-regulated if its numbers always decrease when they exceed a certain threshold K, and always increase otherwise. A simple way to test for this is to arrange the censuses in descending order: under each census place a D if the population decreased between that year and the next, and an I otherwise. A clean separation between D's for higher censuses and I's for lower implies density regulation. The results of this test are as follows:

1500	1200	700	650	600	575	500	425	400	390
D	*D*	*IDD*	*D*	*D*	*D*	*DD*	*D*	*DIII*	*D*
350	320	300	250	200	180	150	140	120	100
D	*D*	*I*	*D*	*I*	*D*	*III*	*I*	*D*	*IIIID*
90	80	75	70	60	50	35	25	20	
I	*II*	*I*	*II*	*II*	*I*	*I*	*I*	*I*	

No time lag is involved if the change in numbers from one year to the next is governed by the geometric means of the censuses in those years. If we plot the increase between years n and $n + 1$ against the geometric means of the censuses in those years, we find a most remarkable confusion of I's and D's, suggesting that this population's regulation does exhibit a time lag. To find the extent of the time lag, one would ask what year's census best predicts the population's change between years n and $n+1$. Plotting increase or decrease between years n and $n+1$ under the censuses of year $n-1$, we find a separation better than if we assume no time lag, but worse than in the first part of this problem.

It should be pointed out that a population participating in a predator-prey oscillation would appear to be density-regulated, with a time lag in the regulation.

4. Suppose *Paramecium aurelia* and *Paramecium caudatum* are competing in a laboratory culture. Let N_1 and N_2 be the numbers of *aurelia* and *caudatum* per cubic centimeter, and let their equations of competition be

$$dN_1/dt = 1.024N_1 - .00535N_1^2 - .00875N_1N_2,$$

$$dN_2/dt = .694N_2 - .00375N_1N_2 - .00612N_2^2,$$

where time is measured in days (data from Gause).

a) Are these species competing for the same niche?

b) Which will win the competition?

Solution: Notice that an additional *caudatum* per cubic centimeter would decrease *aurelia's* growth rate by .00875, while an additional *aurelia* per cm^3 decreases *aurelia*'s growth rate by .00535: one *caudatum* has the same effect as 875/535, or 1.635, *aurelia*. Similarly, we find that one *caudatum* has the same effect on *caudatum*'s growth rate as 612/375, or 1.632 *aurelia*. These two species appear to be sharing a niche where crowding is measured by *aurelia*'s density plus 1.634 times *caudatum*'s density. We may describe this competition fairly accurately by the equations

$$dN_1/dt = 1.024N_1 - .00535(N_1 + 1.634N_2)N_1,$$

$$dN_2/dt = .694N_2 - .00375(N_1 + 1.634N_2)N_2.$$

Aurelia is in equilibrium when $N_1 + 1.634N_2 = 1.024/.00535$, or 191: *caudatum* is in equilibrium when $N_1 + 1.634N_2 = 186$. Thus *aurelia* wins.

5. Consider a mite population limited by a predator which feeds on nothing else. Suppose the mites are sprayed by an insecticide to which they are more sensitive than their predators. Will this reduce the mite population?

Solution: Let N_1 and N_2 be the numbers per acre of mites and their predators. Suppose that, in the absence of insecticide,

$$d \log N_1/dt = e_1 - a_{12}N_2;$$

$$d \log N_2/dt = -e_2 + a_{21}N_1,$$

while, under application of the insecticide,

$$d \log N_1/dt = e_1 - m_1 - a_{12}N_2;$$

$$d \log N_2/dt = -e_2 - m_2 + a_{21}N_1.$$

Here, e_1 is the prey growth rate in the absence of predators or insecticide, and $a_{12}N_2$ and m_2 are their death rates from predators and insecticide, respectively. e_2 is the predator death rate when prey are lacking, $a_{21}N_1$ the gain from the prey, and m_2 the predator's death rate from insecticide. If m_1 is less than e_1 for any genotype among these mites, the prey abundance is destined to increase, because the number of prey required to maintain the predator population is now $(e_2 + m_2)/a_{21}$, whereas before the insecticide was applied this number was e_2/a_{21}. The insecticide lowers the prey's rate of increase, and temporarily thins their number, but their eventual abundance is determined by the efficiency of their predators, and this the insecticide cannot help but lower.

6. Consider the following protocol from a predator-prey experiment (N_1 and N_2 are numbers of prey and predator, respectively):

day	1	2	3	4	5	6	7	8	9	10
N_1	160	37	18	15	30	65	100	110	40	20
N_2	80	170	120	50	15	13	25	60	125	60

day	11	12	13	14	15	16	17	18	19
N_1	15	20	40	75	100	125	125	50	25
N_2	35	20	15	12	22	37	100	160	80

(read from curve of Allee, Emerson, Park, Park, and Schmidt, p. 327, smoothed from data of Gause). What are the isoclines of this interaction?

Solution: Volterra's theory predicts that the predators increase if the prey exceed a certain abundance and decrease otherwise. Likewise, the prey should decrease if the predators exceed a certain threshold. In this experiment, the predators always increase from day n to day $n + 1$ if there are more than 62 prey on day n, and decrease otherwise. The prey always decrease if there are more than 55 predators, and, with one exception, they always increase otherwise. Notice that there were 125 prey on both days 16 and 17: the prey did not increase even though there were only 37 predators on day 16. The isoclines $N_1 = 62$ and $N_2 = 55$ thus fit the data fairly well, although for high numbers of prey crowding effects force the prey isocline downward.

Rosenzweig asserts that, as a rule, the prey are unusually sensitive to predation when their numbers are low (Fig. 7-5).

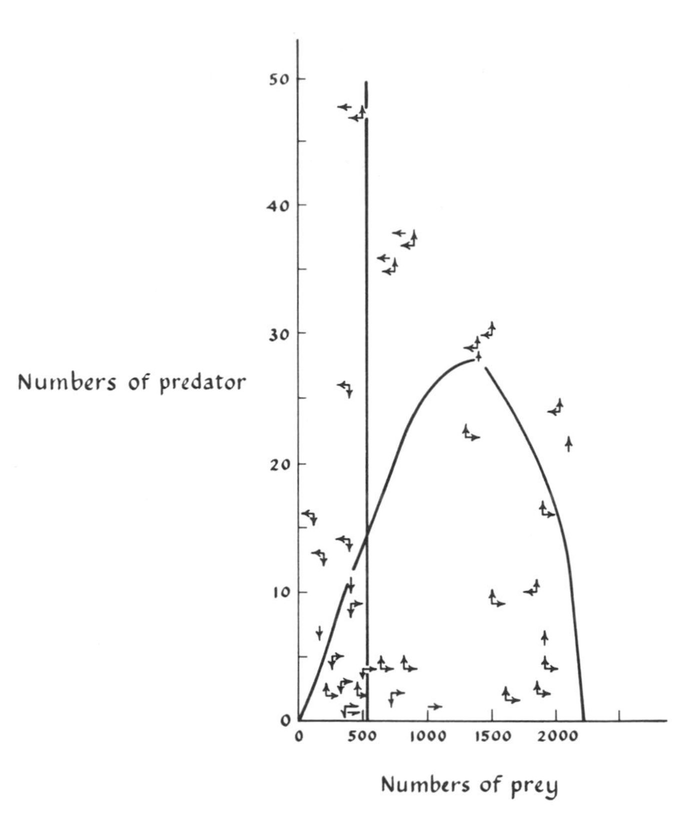

Fig. 7-5. Isoclines for the interaction of a predatory mite and its prey in an experiment by Huffaker. The hump is the prey isocline, the set of points at which the prey population does not change, and the vertical line is the predator isocline. Each point represents one census of predator and prey, based on a *smoothed* version of Huffaker's data: an arrow points upward if the predators increased between this census and the next, and downward if they decreased; the predator isocline separates upward from downward arrows as cleanly as possible. Moreover, a point has an arrow pointing to the right if the prey population were increasing, and to the left if decreasing; the prey isocline is meant to

Bibliographical Notes

The best brief summaries of ecology are G. E. Hutchinson, "Homage to Santa Rosalia, or Why Are There So Many Kinds of Animals?", pp.145-159 of the *American Naturalist*, vol. 93, 1959, and a miniature book by Charles Elton, *The Ecology of Animals* (third edition, Methuen, 1950). Another of Elton's books, *Animal Ecology* (Sidgwick and Jackson, 1927), transformed ecology into a coherent discipline; it is, moreover, one of the most readable of the great books of biology. An outstanding longer work is Kenneth Watt's *Ecology and Resource Management* (McGraw-Hill, 1968). Since Watt needs ecological principles capable of practical application, he avoids the Scylla and Charybdis of pointless description and meaningless generality. W. Hazen has collected many important papers in his volume, *Readings in Population and Community Ecology* (second edition, Saunders, 1970). The classic encyclopedia of the subject is Allee, Emerson, Park, Park, and Schmidt, *Principles of Animal Ecology* (Saunders, 1949);

---

separate leftward arrows from those pointing to the right. Perfect segregation is impossible because the numbers of predator and prey do not completely determine how these numbers will change. (From Fig. 1 of M. Rosenzweig, "Why the Prey Curve has a Hump," p. 82 of the *American Naturalist*, vol. 103, 1969. Copyright 1969 by the University of Chicago; all rights reserved. Used by permission of the University of Chicago Press.)

The founding classics of mathematical ecology are A. J. Lotka, *Elements of Physical Biology*, Williams and Wilkins, 1925 (reprinted by Dover in 1956 as *Elements of Mathematical Biology*), and V. Volterra, *Leçons sur la Théorie Mathématique de la Lutte pour la Vie* (Gauthier-Villars, 1931). Lotka sought to frame a "physical biology" analogous to physical chemistry, and was something of an amateur philosopher besides. Volterra simply frames a mathematics of population interactions, and offers much less to non-mathematical readers. Except for Sir Ronald Fisher, Volterra was the most gifted mathematician to enter biology: his work is guided by an aesthetic sense which substitutes remarkably well (though not completely) for biological knowledge. Some feel that the ideas of mathematical ecology bear no relation to reality: see, for example, Andrewartha and Birch, *The Distribution and Abundance of Animals* (University of Chicago Press, 1954).

The principle of competitive exclusion and its consequences are splendidly summarized by G. E. Hutchinson in *The Ecological Theater and the Evolutionary Play*, Yale University Press, 1965. Experimental studies of competition were first reported by G. F. Gause in *The Struggle for Existence*, Williams and Wilkins, 1934.

Mechanisms of population regulation are discussed by Slobodkin in *Growth and Regulation of Animal Populations*, Holt, Rinehart and Winston, 1961. In his paper, "Random Dispersal in Theoretical Populations," pp. 196-218 of *Biometrika*, vol. 38, 1951, Skellam discusses conditions permitting opportunistic and equilibrium species to coexist. In their paper, "On Time Lags in Equations of Growth," pp. 699-702 in the *Proceedings of the National Academy of Sciences*, vol. 42, 1956, Wangersky and Cunningham discuss conditions causing space-limited populations to oscillate. C.

Huffaker discusses a predator-prey oscillation he created in the laboratory in "Experimental Studies on Predation: Dispersion Factors and Predator-Prey Oscillations," pp. 343-383 of *Hilgardia*, vol. 27, 1957. A simple laboratory oscillation is discussed in J. Drake et al., "The Food Chain," pp. 87-95 in J. F. Saunders, ed., *Bioregenerative Systems*, published by the National Aeronautics and Space Administration, 1966. Oscillations between lemmings and vegetation are discussed in pp. 212-217 of D. Lack, *The Natural Regulation of Animal Numbers*, Clarendon Press, 1954: evidence that overabundant lemmings "crash" when they have eaten all their forage may be found in F. Pitelka, "Some Aspects of Population Structure in the Short-term Cycle of the Brown Lemming in Northern Alaska," pp. 237-251 in the *Cold Spring Harbor Symposia for Quantitative Biology*, vol. 22, 1957. A. Novick discusses the evolution of nutrient-limited bacteria in a chemostat in his paper, "Some Chemical Bases for Evolution of Micro-organisms," in A. Buzzati-Traverso, *Perspectives in Marine Biology*, University of California Press, 1958. M. S. Bartlett provides a summary of these topics suitable for mathematicians in his brief but stimulating monograph, *Stochastic Population Models in Ecology and Epidemiology*, Methuen, 1960.

Hairston, Smith, and Slobodkin's paper is "Community Structure, Population Control, and Evolution," pp. 421-425 of the *American Naturalist*, vol. 94, 1960. Perhaps the best summary of opposing views is P. Ehrlich and L. C. Birch, "The 'Balance of Nature' and 'Population Control,' " pp. 97-107 of the *American Naturalist*, vol. 101, 1967, which emphasizes the difficulty of interpreting such statements as "plants are limited by energy rather than by herbivores." I feel that a plant community is herbivore-limited only if removal of herbivores of all sorts (insects, deer, etc.) causes an increase of over 15% in the weight of foliage per acre, or if it causes great changes in the nature of the vegetation (transforming grassland into savanna or woodland, etc.). This definition is difficult even in theory: how can we remove the herbivores without interrupting pollination and dispersal? It is suggestive, however, that even in Long Island's rather depauperate forest, less than 9% of the leaf surface is eaten while green (see the excellent and instructive paper by R. H. Whittaker and G. Woodwell, "Structure, Production and Diversity of the Oak-Pine Forest at Brookhaven, New York," pp. 157-176 in the *Journal of Ecology*, vol. 57, 1969. In a Puerto Rico rain forest, insects eat about 7% of the leaf surface (H. T. Odum and R. Pigeon, eds., *A Tropical Rain Forest*, AEC Division of Technical Information, 1971). I am informed that the IBP found that insects eat between 0.7% and 5% of the leaves in British forests. Early successions are more eaten, presumably because their plants are too ephemeral to merit elaborate defense.

The facts behind the oft-told Kaibab deer story are now being vigorously questioned: see the discussion in G. Caughley, "Eruption of Ungulate Populations, with Emphasis on Himalayan Thar in New Zealand," pp. 53-73 in *Ecology*, vol. 51, 1970.

It is sometimes easier to learn what limits animal populations: one approach is described in J. Connell, "The Influence of Inter-specific Competition and Other Factors on the Distribution of the Barnacle, *Chthalamus stellatus,*" pp. 710-723 in *Ecology*, vol. 42, 1961.

CHAPTER 8

The Origin of Species

HOW ARE species formed? Imagine an animal which in ancient times learnt to exploit a hitherto unoccupied role, and which thereby spread over all of North America. Suppose that the Rocky Mountains now arise, separating the species into an Eastern and a Western population; suppose, moreover, that the animals cannot cross the mountains, so that the two populations do not interbreed, and evolve separately. Either by chance, or in response to differing environments, the populations diverge. If members of one population somehow cross the mountains, they may establish themselves, or they may be less fit in every respect than the natives to live there, and die out. If they do establish themselves, and if the two populations differ enough in their habits and preferences that they do not interbreed, they will behave as separate species, which in fact they have become. Selection will minimize competition, leading to character displacement: the two species will, in effect, split the niche between them.

What if the populations interbreed? A gene's message is meaningful only in certain contexts. Conditions in the organism must be such that the enzyme our gene specifies functions only in the proper places, at the proper times. These conditions are determined by the other genes the organism carries: they determine the context in which our gene is read. If the two populations are so different that some of the genes of the one are meaningless in the other, hybrids will be less viable or less fertile than their parents. Those individuals which mate with members of their own population will contribute more to the genetics of future generations than those which interbreed with the other. However, individuals refusing to mate with members of the other population may have more difficulty finding mates. If the difficulty of finding mates outweighs the disadvantage of interbreeding, the two populations will fuse. If not, selection will suppress interbreeding between the two populations, and they will become two separate species. In sum, for one species to divide in two, the ancestral population must somehow split, and the two halves must come together again and find some way to coexist.

Darwin's finches, so named because they helped persuade Darwin of the truth of evolution, partitioned a niche in this way. These finches (Fig. 8-1) live on an archipelago, the Galápagos, which lies six hundred miles off the coast of Ecuador. They derive from a common ancestor which presumably reached one of the islands several million years ago. The ancestor and its descendants, finding the avian niche unoccupied, spread over the islands. These islands are quite separated, often thirty or even sixty miles apart. Their bird populations were thus quite isolated, and evolved in different directions. Every once in a while, however, birds must have been blown from one island to another. Often the newcomers were fitted for occupations differing somewhat from that of the natives: perhaps one would prefer insects and the other seeds; perhaps one ate larger seeds than the other; perhaps one preferred coastal mangroves and another drier uplands. The niche was thereby split according to food and habitat preference: where one bird ate both seeds and insects, now one prefers seeds and another insects. As time passed, the finches became remarkably diverse, including fruit-eaters, seed-eaters, and insect-eaters. These birds are extraordinarily instructive. Not only do they illustrate the principles of speciation and character displacement: they show how the ancestral finch make-up has been modified to suit different ways of life. The birds eating the smaller or softer seeds have the smaller bills; the one that seeks insects in the crannies or tree bark finds a cactus thorn with which to pry them out.

Evolution involves keeping up with environmental change and opening up new ways of life, as well as partitioning preexisting niches. What groups are most likely to evolve by partitioning a niche? An individual family or genus is most likely to evolve thus: although birds have evolved for the most part by dividing up a single niche, such large groups usually open up new ways of life as they evolve. Mammals, for example, evolved swimming and flying forms. Moreover, it is only possible to partition *stable* niches: this type of evolution is characteristic of equilibrium species.

How can we tell whether a group evolved by partitioning a niche? MacArthur proceeds by relating the mode of evolution of a group to the relative abundances of its species. He shows that if a group of species divides up a niche in a random way, the relative abundances will approximate a law called the broken stick distribution. Imagine the niche as a stick. If the niche is being divided among n species, choose $n - 1$ points at random on the stick, break the stick at these points, and measure the lengths of the resulting fragments. MacArthur asserts that if we arrange the n species in decreasing order of abundance, the numbers of the jth species will be proportional to the jth longest fragment of the broken stick. This argument is quite contro-

1
2
3
4
5
6
7
8
9
10
11
12
13
14

versial; MacArthur no longer believes it, because it overlooks so many seemingly essential aspects of speciation. However, its predictions do seem to work.

What does the "broken stick" model tell us of the relative abundances of our n species? Since the stick is broken at ran-

---

Fig. 8-1. Species of Darwin's Finches

1.	*Geospiza magnirostris* Gould	Large ground-finch
2.	*Geospiza fortis* Gould	Medium ground-finch
3.	*Geospiza fuliginosa* Gould	Small ground-finch
4.	*Geospiza difficilis* Sharpe	Sharp-beaked ground-finch
5.	*Geospiza scandens* (Gould)	Cactus ground-finch
6.	*Geospiza conirostris* Ridgway	Large cactus ground-finch
7.	*Camarhynchus crassirostris* Gould	Vegetarian tree-finch
8.	*Camarhynchus psittacula* Gould	Large insectivorous tree-finch
9.	*Camarhynchus pauper* Ridgway	Large insectivorous tree-finch on Charles
10.	*Camarhynchus parvulus* (Gould)	Small insectivorous tree-finch
11.	*Camarhynchus pallidus* (Sclater and Salvin)	Woodpecker-finch
12.	*Camarhynchus heliobates* (Snodgrass and Hiller)	Mangrove-finch
13.	*Certhidea olivacea* Gould	Warbler-finch
14.	*Pinaroloxias inornata* (Gould)	Cocos-finch

(After Fig. 3 of D. Lack, *Darwin's Finches*, p. 19, 1947, Cambridge University Press, Cambridge, England, with permission of author and publisher. Reprinted 1961 by Harper Torchbooks, Harper and Row, New York.)

dom, we cannot tell just how long the jth fragment will be, any more than we can tell how many tosses come up heads if we toss a coin a hundred times. However, if we toss a coin a hundred times, we usually get *about* fifty heads, and we will more likely get fifty heads than any other number. We may likewise calculate the most probable length of the jth longest fragment of our stick: if we have broken our stick into many fragments, the actual length will usually approximate the most probable. This means that, even when MacArthur's model applies, the broken stick distribution will be obeyed only if the niche has been partitioned among a relatively large number of species. A group of closely related species with dispersal rates high enough to maintain many representatives in a community is thus the most likely to obey this distribution.

MacArthur predicts that if a niche is divided at random among n species, the relative abundances will be as follows:

most common	$1 + 1/2 + 1/3 + \ldots + 1/(n-1) + 1/n$
next most common	$1/2 + 1/3 + \ldots + 1/(n-1) + 1/n$
next rarest	$1/(n-1) + 1/n$
rarest	$1/n$

In the many communities studied, birds satisfy this law. The nine species of serpent and basket stars which live together in an abandoned intertidal quarry of a South Pacific atoll fit this distribution, as do the fishes of the perch family living together in the one stream studied. The distribution has been checked for two genera of predatory snails, *Conus*, which kills its prey with a paralyzing poison, as the rattlesnake does, and *Morula*, which drills into its prey and sucks out the juice; the species of

Conus living together on a reef often (Fig. 8-2) fit, as do the *Morula*. On the other hand, if one considers together all the predatory snails in a community, their relative abundances do not obey this distribution, for the predatory snails in question do not all derive from a common ancestor. Soil arthropods (Fig. 8-3) do not obey it, for the same reason. Ciliate protozoans, and nematodes, do not fit: most of these are opportunistic species.

Problem

On one coral bench in Hawaii, Kohn collected 231 *Conus ebraeus*, 149 *Conus abbreviatus*, 66 *Conus sponsalis*, 29 *Conus chaldeus*, 10 *Conus lividus*, 7 *Conus flavidus*, 6

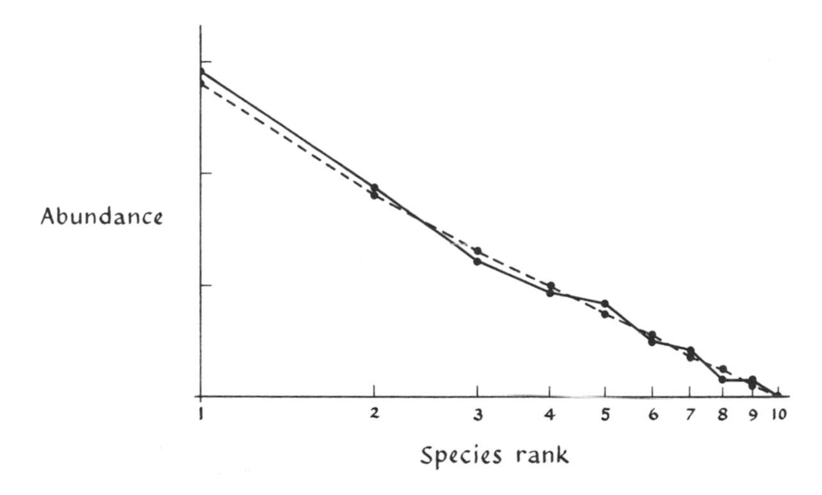

Fig. 8-2. Relative abundances of the species of *Conus* on a Hawaiian reef platform, based on a collection of 182 individuals. (After Fig. 16 of A. J. Kohn, "Ecology of *Conus* on Hawaii," *Ecological Monographs*, p. 58, vol. 29, 1959, with permission of the author and of Duke University Press.)

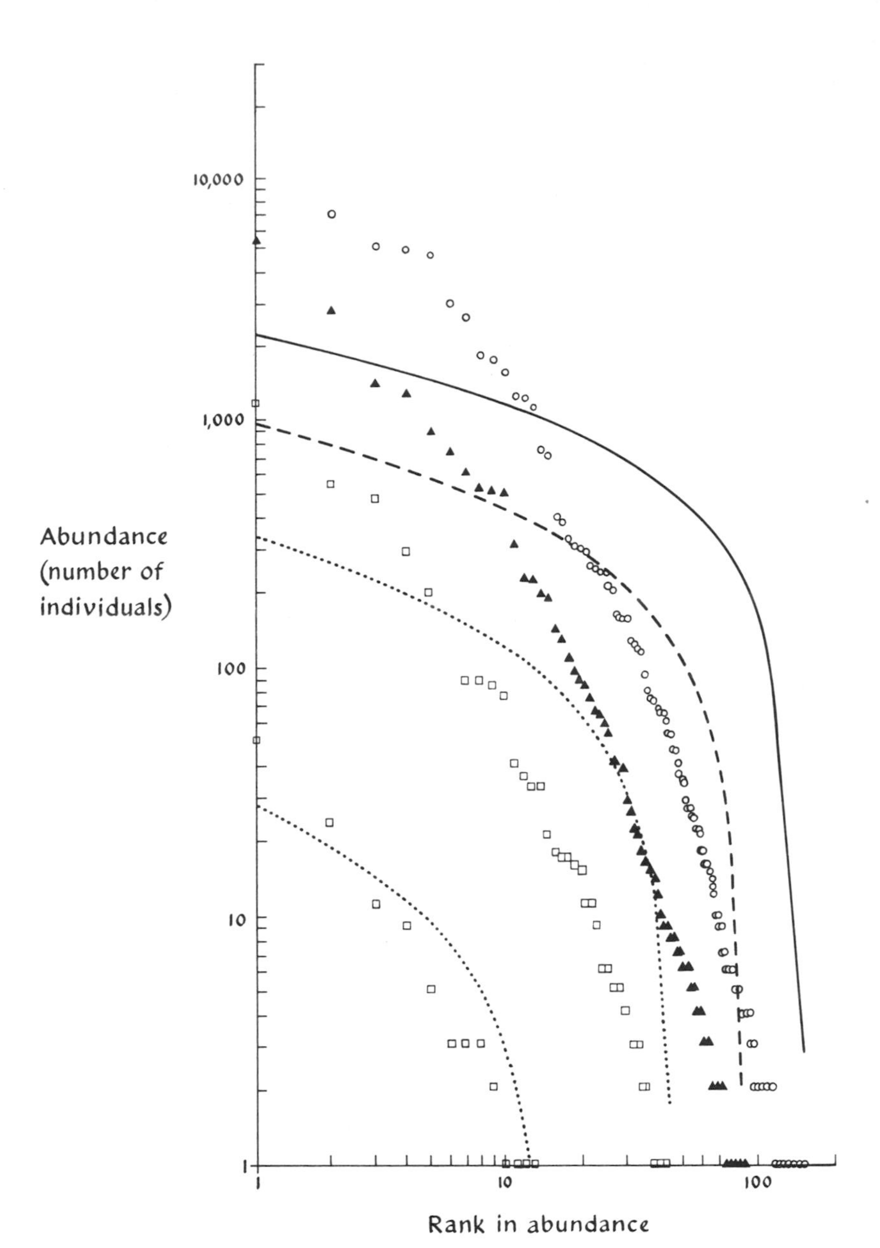

10,000
1,000
100
10
1
Abundance (number of individuals)
10
100
Rank in abundance
LEGEND
Expected
Observed
24 samples lumped
5 samples lumped
single sample (a typical one and the smallest one are shown)

Conus rattus, and 3 *Conus pennaceus* (data read off from a graph).

a) Have these eight species partitioned their niche at random?

b) *Flavidus*, *rattus*, and *pennaceus* are primarily subtidal animals, occurring below the bench. Supposing the individuals of these species were outside their normal habitat, can we say the remaining five have partitioned their niche at random?

Solution: If all eight species have partitioned a niche at random, their relative abundances should be:

$$
\begin{array}{r}
1 + 1/2 + 1/3 + 1/4 + 1/5 + 1/6 + 1/7 + 1/8 \\
1/2 + 1/3 + 1/4 + 1/5 + 1/6 + 1/7 + 1/8 \\
1/3 + 1/4 + 1/5 + 1/6 + 1/7 + 1/8 \\
1/4 + 1/5 + 1/6 + 1/7 + 1/8 \\
1/5 + 1/6 + 1/7 + 1/8 \\
1/6 + 1/7 + 1/8 \\
1/7 + 1/8 \\
1/8
\end{array}
$$

Thus 1/64 of the individuals collected should belong to the rarest species, 1/64 + 1/56 to the next rarest, 1/64 + 1/56 + 1/48 to the third, etc. Since 500 individuals were collected in all, the numbers collected, starting with the rarest species, should have been 8, 17, 27, 40, 55, 76, 107, and

Fig. 8-3. Relative abundances of soil arthropods in samples of various sizes, compared with predictions from MacArthur's distribution. Data from Hairston and Byers. (After Fig. 5 of N. G. Hairston, "Species Abundance and Community Organization," *Ecology*, p. 412, vol. 40, 1959. Used with permission of the author and of Duke University Press.)

170, respectively. In fact the common species are more common, and the rare species much rarer, than the theory suggests.

If we exclude the rarest three species as immigrants, the relative abundances of the other five should be:

$$
\begin{array}{r}
1 + 1/2 + 1/3 + 1/4 + 1/5 \\
1/2 + 1/3 + 1/4 + 1/5 \\
1/3 + 1/4 + 1/5 \\
1/4 + 1/5 \\
1/5
\end{array}
$$

Thus 1/25 of the individuals collected should belong to the rarest species, 1/25 + 1/20 to the next rarest, etc. As 484 individuals were collected of the five species which "belonged" there, the numbers of each species, starting with the rarest, should have been 19, 44, 76, 124, and 221. In fact, they were 10, 29, 66, 149, and 231. We have no objective means of deciding whether the theory's prediction is verified by the data, but it would seem this is the case.

Bibliographical Notes

There is an enormous literature on the origin of species. Two outstanding books are Th. Dobzhansky, *Genetics and the Origin of Species*, Columbia University Press (second edition 1941, third 1951), and E. Mayr, *Systematics and the Origin of Species*, Columbia University Press, 1942. A more exhaustive summary is E. Mayr, *Animal Species and Evolution*, Harvard University Press, 1963. G. Stebbins gives a brief and readable survey of the subject in his book, *Processes of Organic Evolution*, Prentice-Hall, 1966.

The significance of Darwin's finches is beautifully explained by David Lack in *Darwin's Finches*, Cambridge University Press, 1947 (reprinted by Harper Torchbooks, 1961).

The "broken stick" model is derived in R. MacArthur, "On the Relative Abundance of Species," pp. 25-36 of the *American Naturalist*, vol. 94, 1960. Relevant data are presented in his paper and in A. J. Kohn, "The Ecology of *Conus* in Hawaii," pp. 47-90 of the *Ecological Monographs*, vol. 29, 1959; N. Hairston, "Species Abundance and Community Organization," pp. 404-416 of *Ecology*, vol. 40, 1960; and C. E. King, "Relative Abundance of Species and MacArthur's Model," pp. 716-727 of *Ecology*, vol. 45, 1964.

Hairston believes that a fit to the MacArthur distribution is an artefact of sample size (see his article in *Ecology*, vol. 50, 1970): only one student that I know of, Dr. Goulden, of the Academy of Natural Sciences of Philadelphia, has sound evidence of fits which are not artefacts of this kind (see Chapter 11).

CHAPTER 9

Plant Communities

WE NOW turn to the patterns in the interactions among plants and animals that live together: we are concerned, in other words, with the organization of a community. To learn how a community is organized, we must ask how it evolves. The fitness of an organism for its way of life is a consequencc of its relation to its parents. The form and structure of an organism is programmed in its genes, which are copies of those of its parents. The copying of genes sometimes leads to misprints affecting the fitness of their bearers. It is an old story how the fitness of an organism results from the age-old, quite automatic, quite inevitable sifting of such misprints according to their effect on reproductivity. However, one cannot often distinguish a *community* from its neighbors (one forest type grades imperceptibly into another, etc.) and one certainly cannot pick out the parents of a community: it is thus meaningless to speak of a natural selection of communities. In a newly dug pond or a newly formed volcanic island, we may watch a community form by a continual colonization. Light seeds, and perhaps stray insects, drift in on the wind; a stray bird lands with heavier seeds in its gut and perhaps small snails on its toes. A very rare log with ants or even a larger animal on it might reach our island; a tornado might bring fish into the pond. A slow but continuous parade of living things drifts or wanders by, as the case may be, any one of which might settle if it found an opportunity. In short, a community is formed through invasion and colonization from other communities. The community organization that persists is that which is most resistant to invasion from the outside. The experience of the last few hundred years, during which man has introduced so many species to new environments, is that complex natural communities are most resistant to invaders: it is the roadside and the railway bank, the garden and the farm, which provide homes for foreign plants, not the native forest.

The species of a community usually interact by crowding or eating one another. In this chapter, we shall treat space-limited communities, and in the next, communities where only food-web limitations are important. The one should give us some understanding of plant, the other of animal ecology.

How do plants interact? It is often assumed they do so primarily by shading each other, but this is not always true. One tree may shade another and yet die for lack of water, a victim of root competition. In the creosote and sage deserts of Utah, a disproportionate abundance of plants grow just off the road, where they catch the runoff when it rains: it is not an increase of light which allows them to grow so densely together. A closer look at the desert shrubbery reveals a remarkably even spacing of the bushes, closer together in the channels where rainwater travels, farther apart elsewhere. The camel thorns and succulents of the Egyptian desert are also distributed thus: the spacing of these bushes can only reflect competition for water.

Water and light must be equally important to a plant. Light supplies the energy for photosynthesis, but that energy is used, in first instance, to split water molecules. Moreover, the capillary action by which water molecules evaporating at the surface draw others behind them, bringing up water from the roots, is a basic transport mechanism of plants. The supply of water, like that of sunlight, is fixed: plants cannot extinguish it. Competition for water must therefore take the form of crowding: if two populations were competing only for water, we could measure each population's influence by its water uptake, just as we measure the populations of plants competing only for light by the shade they cast.*

The water supply and the availability of light each limits the abundance of plants. Selection always acts to increase the lower of the two limits, even at the expense of decreasing the higher: plants therefore evolve so that, in their normal environment, the limits set by water and light are very nearly the same. Any advantage in competition for water permits increased growth and more effective competition for light: we may thus view competition for water as an aid to competition for light.

What are the consequences of competition for light? Recall the equations for two species with identical preferences in growing conditions which compete only for light. If N_1 is the population size (judged by the shade cast) of the first species and N_2 that of the second, then

$$d \log N_1/dt = a_1(K_1 - N_1 - N_2);$$
$$d \log N_2/dt = a_2(K_2 - N_1 - N_2).$$

If K_1 exceeds K_2, the first species can tolerate shadier conditions than the second, and wins the competition. There will be enough light in an equilibrium stand of the second to permit the

* We don't know how plants compete for water. This means we don't know how a plant could increase its water uptake, whereas it is easy to see how light intake could be increased by redistributing leaves. Thus we would be unable to test a theory of competition for water, whereas the theory of light competition is easy to test.

first to invade, but an equilibrium stand of the first will be too shady to permit the second to grow. We derive this from the equations by observing that when the first species invades an equilibrium stand of the second, N_1 is nearly 0, N_2 nearly K_2, and $d \log N_1/dt$ is positive. On the other hand, if the second species invades an equilibrium stand of the first, N_2 is nearly 0, N_1 nearly K_1, and $d \log N_2/dt$ is nearly $a_2(K_2 - K_1)$, which is negative.

On the Eastern seaboard, most trees have either broad, thin leaves which they shed every year, or waxy, stiff, needle-like leaves which they keep all year around. If a deciduous and an evergreen compete, which will win? The broad leaves of the deciduous are more efficient than the needles of the evergreens, but they are more easily damaged by frost. Moreover, the broad leaves evaporate water at a higher rate, a serious disadvantage in the prolonged droughts winter causes by freezing water in the soil. If the increased efficiency of deciduous leaves pays the "overhead expenses" of making new leaves every year, the deciduous can crowd out the evergreens; otherwise, the evergreens win. Deciduous leaves pay in Pennsylvania and California, where the growing season is mild; the harsher conditions of the far north favor evergreens. On the other hand, tropical trees keep broad, thin leaves the year around: their growing season never stops. Local distributions also tally with this story. In the canyons of the Rockies and the Sierra, evergreens grow up the canyon walls and deciduous trees grow along the river, where there is enough water to justify making new leaves every year. In the ravines of eastern Pennsylvania, the length of the growing season limits tree growth more than rainfall does. Here, hemlocks grow in the shadowy ravine bottoms where the frost lies, and deciduous trees grow up the sunnier slopes.

Deciduous and evergreens sometimes coexist. In Colombia the foothills of a great mountain range, the Sierra Nevada de Santa Marta, extend to the Caribbean. The coastal hills receive most of their rain in the fall, and occasional storms at other times. They are covered by so-called "dry forest" consisting of scrubby trees ten or twenty feet high, varied by occasional candelabra cacti. In some places deciduous and evergreens mingle in a remarkably uniform manner. In March, these foothills are extraordinarily beautiful: one sees patches of orange flowers representing a legume bearing long seedpods resembling those of the honey-locust, and a pattern of greens, some dark, some bright, against a brushy background of dry and leafless branches. Mixtures of deciduous and evergreens also occur in forests and chaparral of coastal California: the evergreens have thick and leathery leaves, often rather hairy; those of the deciduous plants are thinner and more delicate. What permits deciduous and evergreen foliage to coexist? The evergreens face no competition during the dry season, and more effective competition during the rainy season. If enough rain falls "out of season" to permit the evergreens to recover from crowding during the rainy season, the two will coexist.

In general, two competing species will coexist if they prefer somewhat different growing conditions. If each species can invade an equilibrium stand of the other, then neither can quite crowd out the other and the two will coexist. If, for example, beeches prefer lower, wetter places and oaks the drier hilltops, an oak can invade an equilibrium stand of beeches if there is enough light on the hilltops to permit it to grow. Even though beeches cast a far deeper shade in their favorite bottomlands than oaks do on their hilltops, oaks can invade if the beeches so prefer the bottoms that they do not cover the hills too thickly. To phrase this mathematically, let N_1 be the average shade the beeches cast per unit area, and N_2 that cast by the oaks. Assume, moreover, that oak shade has only α times the effect on beech growth that beech shade does, and vice versa. α measures the overlap in the preferences of the two species: if α is zero, they prefer entirely distinct habitats (and thus do not compete at all); if it is one, they prefer the same growing conditions. Then

$$d \log N_1/dt = a_1(K_1 - N_1 - \alpha N_2);$$

$$d \log N_2/dt = a_2(K_2 - N_2 - \alpha N_1).$$

K_1 is the amount of shade the beeches can tolerate and K_2 is the same for the oaks. The oaks can invade an equilibrium stand of beeches if $d \log N_2/dt$ is positive when $N_1 = K_1$ and N_2 is nearly zero: this is true when K_2 exceeds αK_1. Likewise, the beeches can invade an equilibrium stand of oaks if K_1 exceeds αK_2. If K_2 is less than K_1, the condition of coexistence is that α be less than K_2/K_1: if the oaks are much less tolerant of shade than the beeches, their preferences cannot overlap much if the two are to coexist. If the shade tolerances of the two species vary with changing conditions, α must be small enough that one species does not die out while the other is superior.* The less stable the environment, the more distinct the preferences of the competing species must be to permit coexistence.

In the tropical forests of Panama, there are two sizes of tree. The taller trees are 100 to 120 feet high, and spaced far

---

* This theory applies to the coexistence of deciduous and evergreens. Evergreens are in leaf all the while deciduous are, so evergreen shade affects deciduous growth as much as deciduous shade does. On the other hand, deciduous shade has only p times the effect on evergreen growth rate that evergreen shade does, where p is the proportion of the evergreen's photosynthesis that would take place during the rainy season in the absence of deciduous competition. Let N_1 be the evergreen population, and N_2 that of the deciduous. Then

$$d \log N_1/dt = a_1(K_1 - N_1 - pN_2);$$
$$d \log N_2/dt = a_2(K_2 - N_1 - N_2).$$

The conditions for coexistence are $K_2 > K_1$; $K_1 > pK_2$: the deciduous must be more efficient than the evergreen during the rainy season, but the evergreens must have enough "unseasonable" warmth or rain to make up their good season losses.

enough apart so that their crowns rarely touch. Their leaves look very much alike: save for the occasional tree with compound leaves, like those of the acacia or honey-locust, they nearly all have entire-margined leaves of a size and shape intermediate between elm and magnolia (Fig. 9-1). The shorter "understory trees," on the other hand, are close enough together that their crowns often touch to form a canopy. Their leaves assume the exotic sizes and shapes one associates with the tropics: mingled with quite ordinary foliage are palms, tree ferns, and heliconians (bird-of-paradise plants) with great spikes of bizarre red flowers and huge banana-like leaves (Fig. 9-2). It is as if the ground-herbs of a temperate forest, which are so often more varied in leaf form than the trees above them, had become trees in their own right. The tall and short trees are of course adapted to different degrees of shade, but there is undoubtedly enough competition that a slight thinning out of one kind would permit more seedlings of the other to grow. The shorter trees, however, need some tall ones to protect them from the full force of the tropical sun: the tall trees make it possible for other plants to specialize to shady conditions. A need for shade is a familiar matter to the horticulturist: the May-apple, which forms a carpet in many Eastern forests, will also grow in the shade of houses, and many strains of coffee grow best under shade trees.

Plants not only take light and water from each other: they sometimes poison each other. Californians may be familiar with the eucalyptus tree, which has a characteristic odor deriving from volatile elements in its bark and leaves. It sheds its bark in great strips: a grove of eucalypts makes quite a litter, which inhibits the growth of many plants. Some desert plants poison each other to keep competitors for water at a distance. Even quite familiar plants affect the chemistry of their soil. The collector of land snails does not usually look for them in coniferous forests, for the conifers create an acid soil that dissolves snail shells.

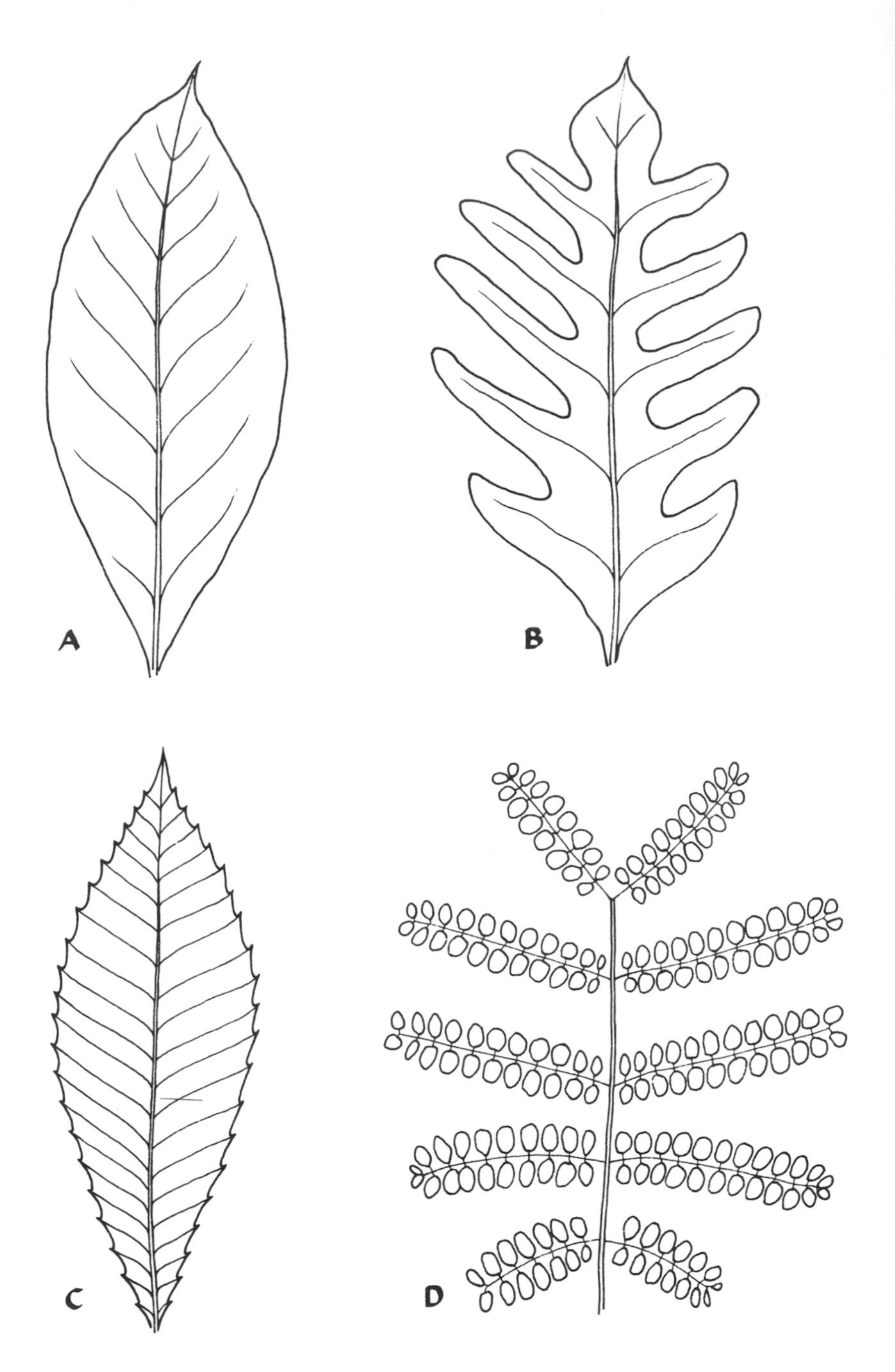

Fig. 9-1. Leaf outlines. A, B, and C are simple leaves; D is compound. A is entire-margined and has a "drip tip," B is lobed, C is serrate.

It is quite likely that plants do not restrict themselves to chemical poisons. The pines of the New Jersey barrens are easily set afire: some of them need fires to open their cones and release the seeds. The fires turn the soil into a leached, barren sand which supports a vegetation far less diverse than that of neighboring regions. In some places, the soil is so poor that the pines can only grow waist high in twenty years. Yet when fires are stopped long enough, a diverse vegetation eventually develops. Have the pines evolved to "encourage" fires which burn some competitors and wreck the soil for the others? Perhaps they have exploited a soil easily wrecked by burning, and are unable to spread beyond it. In tropical South America there is a

Fig. 9-2. A heliconian ("bird-of-paradise" plant).

"white sand forest" restricted, as its name implies, to sandy soil. Like the pine barrens, this community is far less diverse than its neighbors. Perhaps this forest facilitates the leaching of its soil by the torrential tropical rains, to keep out invaders. This forest has been spreading lately; perhaps human disturbance has allowed it to encroach upon new ground.

Consider a tree which poisons its soil against competitors. An equilibrium stand of this tree resists invasion by more fit competitors: the poison makes this stand seem more crowded than it really is. On the other hand, had one of its more fit competitors settled first, the poisonous tree might well find conditions too crowded to invade. Poisons enhance the effects of "accidents of history": if trees compete by poisoning each other, a community is settled by those species which happen to arrive first, whether or not they are the most shade-tolerant. Let N_1 be the shade cast by a poisonous species of tree (immune, of course, to its own poison), and N_2 that cast by a competitor. Assume that the second species is more tolerant of shade than the first, so that $K_2 > K_1$, and that the first species affects the growth rate of the second through the shade it casts and the poison it releases, while the second species only shades the first. Assume, finally, that the poison does not spread enough to allow its makers to encroach upon full-grown competitors. If the poison magnifies the effect of the first species on the second by a factor $1 + z$, then

$$d \log N_1/dt = a_1(K_1 - N_1 - N_2);$$

$$d \log N_2/dt = a_2(K_2 - (1+z)N_1 - N_2).$$

The first species cannot invade an equilibrium stand of the second: if N_1 is nearly 0 and N_2 is K_2, $d \log N_1/dt$ is negative. Likewise, if K_2 is less than $(1+z)K_1$, the second species cannot invade a stand of the first.

The poison which keeps some plants out may provide,

however, quite the optimum environment for others. Many redwood forests favor the growth of carpets of redwood sorrel, a clover-like plant with leaves somewhat acid to the taste and a flower recalling a Spring Beauty. Perhaps the redwoods provide shade and a soil chemistry suiting the sorrels; perhaps, however, the sorrels would grow as well elsewhere were it not for competitors which the redwoods poison. The azaleas of the Atlantic seaboard thrive on the acid soils of the pine barrens, and will not tolerate more alkaline conditions.

Consider two species competing for light, but suppose each benefits by a secretion of the other. These secretions cause each species to be less harmed by increase of the other's population than by increase in its own: mathematically, it is as if the two species differed somewhat in their preferences. In a lake, many micro-organisms appear to live the same way: are beneficial secretions responsible for their diversity? If one fills ten jars with water from a clear mountain stream, adding a few drops of sterile milk to each for food, the protozoans will multiply far faster if water is exchanged among the jars. The only effect of the exchange is to diversify the populations of each jar: apparently the protozoans grow better in the presence of other species. Just how they interact we do not know, nor are we sure other micro-organisms interact in similar ways, but beneficial secretions do seem to be important to protozoans. They may also be important among higher plants. Before the time of chemical fertilizers, a farmer planted nitrogen-fixing legumes to fertilize his soil: perhaps such plants play the same role in nature.

What patterns does the community's evolution impose on these interactions? If plants are space-limited, the plant community most resistant to invasion is that in which each habitat is as crowded as possible. A species specializes to exploit fully some way of life, to crowd fully some habitat, no matter how restricted. Admittedly, it does no good to specialize to a temperature range so narrow that by doing so the species dies of freezing or overheating. To avoid replacement by more specialized competitors, however, a species must specialize as much as changing conditions allow.

If the spectrum of possible habitats is largely the same from community to community, then a stabler environment implies a more diverse community. Environmental stability implies some freedom from unpredictable change. The San Francisco Bay area is relatively stable because the rain there falls fairly evenly, during a specific season; Arizona receives her rain in occasional violent thunderstorms. This makes the difference between a green and lovely land, and a desert. Along California's Redwood Highway, the streams are susceptible to floods of unpredictable violence and extent. The trees on the stream bottomlands are all redwoods, save for the occasional opportunistic tanbark oaks that spring up where one of the giants has fallen; even the undergrowth, where there is any, is monotonous. Out of reach of the floods, the forest diversifies. The redwoods are joined by great Douglas firs and leaning, curving madrones with yellowish-green, magnolia-shaped leaves and somewhat peeling maroon bark. On the other hand, the hurricanes of South Florida, and the annual floods which occurred before canals linked Okeechobee to the sea, seem to have permitted a most remarkable diversity of plants.

Stability also implies relatively uniform conditions. The tropics are stabler than the temperate zone because the range of temperature and wind is less. In spite of their violent hurricanes, the Everglades contain a diversity of trees unknown in those parts of the East Coast subject to even occasional frost. Farther south, in Panama, is an entirely new world, in which buildings need only screens on the windows, because it is never cold and rain blows in so seldom. This is a world in which freezes are unknown and storms quite rare, a world which is accordingly fantastically diverse.

This increase in diversity from the poles to the equator is one of the great generalizations of biology. It is only a correlation, however. Walking along a mountain path in California's Santa Cruz, far out of reach of floods, turning a corner from a shadier to a sunnier hillside can bring one from a uniform stand of redwoods to a remarkably diverse woodland. Such a difference would be entirely unreasonable if the range of possible habitats were similar from community to community. However, we have no reason to think this: the plants themselves create these habitats, forming their peculiar gradients of light, soil chemistry, etc.; and different plants can make very different worlds for their associates.

Since plants compete by shading each other, the mature plant community casts the deepest shade. The mature community's foliage density (the total leaf surface per unit ground surface) should therefore be the maximum the environment permits. Plant communities in similar environments should therefore have similar foliage densities, whatever their other differences might be. For example, a forest and a neighboring grassland should support similar foliage densities. Information still lacks on this score: we have to test the idea in other ways.

A plant derives its energy from photosynthesis: the more photosynthesis it carries on, the better it is able to grow and shade its competitors. Since all the plants in a community are competing for light, the photosynthesis and therefore the productivity of the entire community tend to the maximum the environment permits. Productivity is the rate of growth of new plant matter: new wood, leaves, roots, seeds, and fruit. This growth is measured by its "energy content," the heat it would give off if burned in a bomb calorimeter. Thus productivity is the rate at which the energy supplied by the sun is "fixed" as plant matter.

Is the productivity of a community the maximum its environment permits? Rosenzweig tells us that a community's productivity can be estimated from its "actual evapotranspiration," the amount of water evaporating from the soil or transpiring from the foliage. Evapotranspiration can in turn be estimated from tables of monthly temperature and rainfall: the productivity of a plant community can therefore be predicted from its climate (Fig. 9-3). This only holds, however, for plants which interact mainly by competing for light and water. The pine barrens of New Jersey are less productive than neighboring communities of similar climate: the pines, however, compete by poisoning their competitors. Not many communities are based on poisonous interactions, however, for it pays other plants to exploit unused resources by becoming immune to the poison. The more common the poison, the more plants will adapt to it.

Problems

1. The logistic equation of growth may be written

$$\frac{1}{N^2}\frac{dN}{dt} = \frac{r}{N}\left(\frac{K-N}{K}\right).$$

If we set $u = (K-N)/N$, then

$$\frac{du}{dt} = \frac{d}{dt}(K/N) = -\frac{K}{N^2}\frac{dN}{dt}.$$

We may therefore express the logistic as $du/dt = -ru$. Plotting $\log_e N/(K-N)$ as a function of time, we obtain a straight line with slope r. We may use this relation to find r if we already know K: it avoids the need to calculate $d \log N/dt$ from an empirical record of N.

Setting $K = 217{,}000{,}000$, use this method to find r for the population of the United States between 1790 and 1950, using the data in problem 1 of Chapter 6.

Solution: Form the following table:

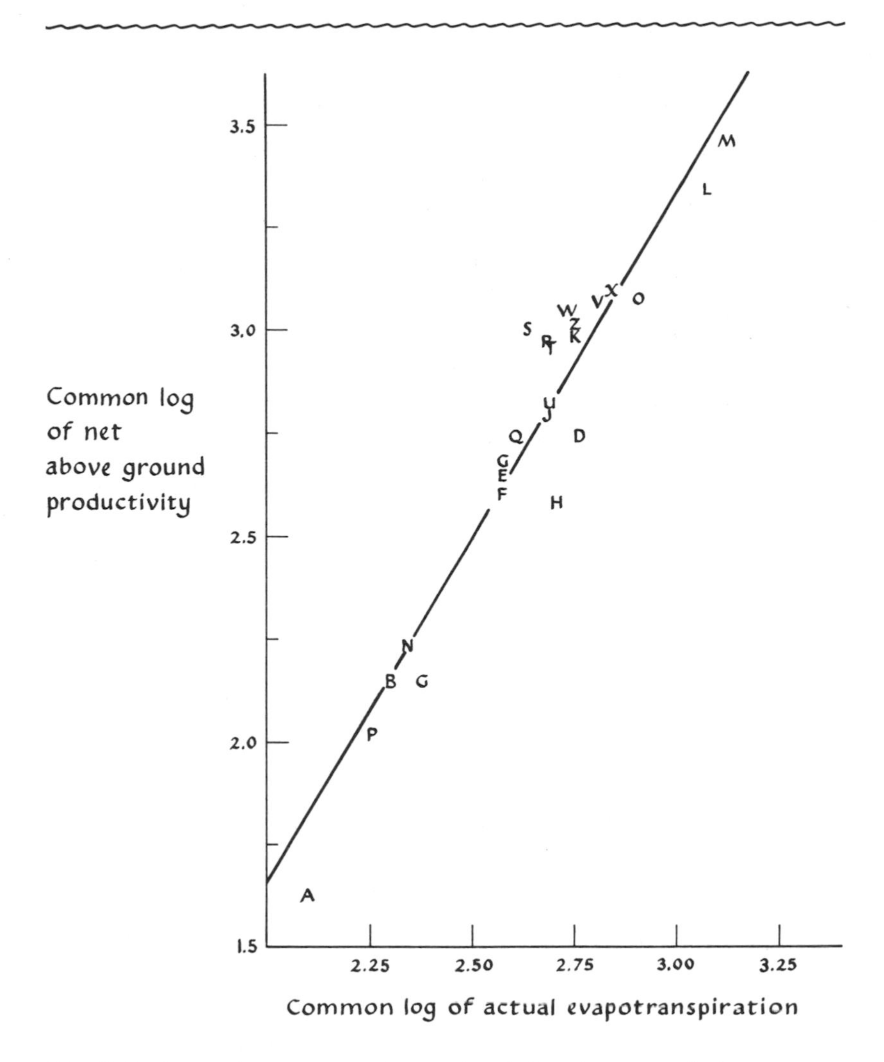

Fig. 9-3. Net above-ground productivity (grams of dry matter per square meter per year) graphed against actual evapotranspiration (millimeters per year). The line best fitting the data is included. (After Fig. 1 of M. Rosenzweig, "Net Primary Productivity of Terrestrial Communities: Prediction from Climatological Data," the *American Naturalist*, p. 72, vol. 102, 1968. Used with permission of the author and of the University of Chicago Press.)

year	N	$K - N$	$\log_{10} N$	$-\log_{10}$ $(K - N)$	$\log N-$ $\log K-N$
1790	3,930,000	213,000,000	6.594	−8.328	−1.734
1810	7,240,000	209,800,000	6,860	−8.322	−1.458
1830	12,870,000	204,100,000	7.110	−8.310	−1.200
1850	23,190,000	193,800,000	7.365	−8.287	−.922
1870	39,820,000	177,200,000	7.600	−8.248	−.648
1890	62,950,000	154,000,000	7.799	−8.178	−.379
1910	91,970,000	125,000,000	7.964	−8.097	−.133
1930	122,780,000	94,200,000	8.089	−7.974	+.115
1950	150,700,000	66,300,000	8.178	−7.822	+.356

These data are adequately summarized by the equation

$$\log_{10} N - \log_{10}(K - N) = .01306(t - 1790) - 1.714,$$

where $t - 1790$ is the number of years elapsed since 1790.

$$r = d \log_e N/dt - d \log_e (K - N)/dt$$
$$= 2.3026 \frac{d}{dt} \log_{10}(N/K - N) = .0301.$$

Thus $r = .0301$.

2. Two experiments growing the yeast *Saccharomyces sp.* in isolation yielded the following protocols (data from Gause):

Age in hours	*Volume of yeast*	*Age in hours*	*Volume of yeast*
0	.45	0	.45
6	1.37	8	1.63
16	8.87	15	6.20
24	10.66	24	10.97
29	12.50	33	12.90
40	13.27	52	12.90
48	12.87		

Find the equation which best describes this yeast's population growth.

Solutions: These data promise no very precise fit to the logistic, so it is best to estimate K and find r from the slope of log $N/K-N$, as in Problem 1. Let $K = 13$. Combining the data of the two experiments, and setting $N(24) = 10.81$, we may form the following table:

t	0	6	8	15	16	24
$\log_{10} N/K-N$	-1.447	$-.928$	$-.845$	$-.041$	$+.332$	$+.693$

We include no later readings, as $\log(K-N)$ is very sensitive to "errors" in N when N is close to K.

Let $u = \log_{10} N/(K-N)$. The simplest way to estimate r is to find the slope of the line connecting $u(0)$ and $u(24)$, and multiply by $\log_e 10$, or 2.303: this is

$$2.303(1/24)[u(24) - u(0)],$$

or .205.

A more reliable way to estimate r is to find the "regression" of $u(t)$ on t, the line $bt - a$ such that the sum of the squared deviations of the "observed values" $u(t)$ from $bt-a$ is as small as possible. In symbols, we seek a and b minimizing the sum

$$S(a, b) = \sum_{i=1}^{6} [bt_i - a - u(t_i)]^2 .$$

Here, t_i are the times of observation, and

$$\sum_{i=1}^{6} x_i$$

is $x_1 + x_2 + \ldots + x_6$. In our case, $S(a, b)$ is

$$(1.447 - a)^2 + (.928 - a + 6b)^2 + (.845 - a + 8b)^2$$
$$+ (15b - a + .041)^2 + (16b - a - .332)^2 + (24b - a - .693)^2.$$

$S(a, b)$ is minimum if $\delta S/\delta a = \delta S/\delta b = 0$. $\delta S/\delta a$ is

$$2(a - 1.447) + 2(a - 6b - .928) + 2(a - 8b - .845)$$
$$+ 2(a - 15b - .041) + 2(a - 16b + .332) + 2(a - 24b + .693).$$

The condition $\delta S/\delta a = 0$ yields the equation

$$1)\ 6a = b(6 + 8 + 15 + 16 + 24)$$
$$- (.693 + .332 - .041 - .845 - .928 - 1.447).$$

Similarly, the equation $\delta S/\delta b = 0$ yields

$$0 = -12(a - 6b - .928) - 16(a - 8b - .845)$$
$$- 30(a - 15b - .041) - 32(a - 16b + .332)$$
$$- 48(a - 24b + .693).$$

Dividing by 2 and rearranging, we obtain the equation

$$2)\ a(6 + 8 + 15 + 16 + 24) = b(6^2 + 8^2 + 15^2 + 16^2 + 24^2)$$
$$- [24(.693) + 16(.332) + 15(-.041) + 8(-.845)$$
$$+ 6(-.928)].$$

Solving equation 1 for a and substituting into equation 2, we obtain

$$\frac{b}{6}(6 + 8 + 15 + 16 + 24)^2$$
$$- \frac{1}{6}(.693+.332-.041-.845-.928-1.447)(6+8+15+16+24)$$
$$= b(6^2 + 8^2 + 15^2 + 16^2 + 24^2)$$
$$- [24(.693) + 16(.332) + 15(-.041) + 8(-.845)$$
$$+ 6(-.928)].$$

We may express b as

$$\frac{\text{Covariance (time } t \text{ of observation, value of } u(t))}{\text{Variance (times of observation)}},$$

where the variance in times of observation is

$$\frac{1}{6}\sum_{i=1}^{6} t_i^2 - \left[\frac{1}{6}\sum_{i=1}^{6} t_i\right]^2$$

and the covariance between the time of observation and the the value of $u(t)$ is

$$\frac{1}{6}\sum_{i=1}^{6} t_i u(t_i) - \left[\frac{1}{6}\sum_{i=1}^{6} t_i\right]\left[\frac{1}{6}\sum_{i=1}^{6} u(t_i)\right].$$

Carrying out the arithmetic, we find $b = .095$, $r = 2.303(.095)$, or .219. The equation of growth for this yeast is therefore

$$d \log_e N/dt = .219(1 - N/13).$$

3. Two experiments growing *Schizosaccharomyces* in isolation yielded the following protocols (data from Gause):

Age in hours	*Volume of yeast*	*Age in hours*	*Volume of yeast*
0	.45	0	.45
16	1.00	15	1.27
29	1.73	32	2.33
48	2.73	52	4.56
72	4.87		
93	5.67		
141	5.83		

Find the equation which best describes this yeast's population growth.

Solution: We must seek the regression of $u(t)$, or $\log_{10} N(t)/(K - N(t))$, on t, using the figures given below:

t	0	15	16	29	32
$u(t)$	−1.078	−.555	−.684	−.375	−.177

t	48	52	72
$u(t)$	−.055	+.555	+.705

The slope of the line connecting $u(0)$ with $u(72)$ is .0248; the slope of the regression line is .0251; r = .058; and

$$d \log_e N/dt = .058(1 - N/583).$$

4. When the two yeasts of Problem 2 and 3 were grown together, their volumes changed as follows (data from Gause):

hour	*Saccharomyces*	*Schizosaccharomyces*
15	3.08	.63
24	5.78	1.22
33	9.47	1.23
44	10.6	1.10
52	9.9	.96

a) Assume these yeasts follow the growth equations

$$\frac{1}{N_1}\frac{dN_1}{dt} = r_1 \left[\frac{K_1 - N_1 - \alpha N_2}{K_1}\right],$$

$$\frac{1}{N_2}\frac{dN_2}{dt} = r_2 \left[\frac{K_2 - N_2 - \beta N_1}{K_2}\right],$$

where N_1 is the population of *Saccharomyces* and N_2 that of *Schizosaccharomyces*. Assume the r's and K's are as determined in Problem 2. What are the values of the competition coefficients α and β?

b) Are these populations regulated by a common factor?

Solutions: Species 2 declines when $K_2 - N_2 - \beta N_1$ is negative. This first happens near hour 33, when $N_2 + \beta N_1$ is nearly equal to K_2, or 5.83. At hour 33 $N_1 = 9.47$, $N_2 = 1.23$, and $\beta = 4.6/9.47$, or .486.

Species 1 declines when $K_1 - N_1 - \alpha N_2$ is negative. This first happens near hour 44, when $N_1 + \alpha N_2$ just exceeds K_1. α is thus 2.4/1.1, or 2.17.

A unit increase in the volume of *Schizosaccharomyces* has α times the effect on the growth rate of *Saccharomyces* as a unit increase of *Saccharomyces* itself, while a unit increase in the volume of *Schizosaccharomyces* has $1/\beta$ the effect on its own growth rate as a corresponding increase in *Saccharomyces*. Since α is 2.17, while $1/\beta$ is 2.06, crowding can be measured for both species by the volume of *Saccharomyces* plus a little over twice the volume of *Schizosaccharomyces*.

Gause calculated α from the formula

$$\alpha = \frac{K_1 - N_1 - (K_1/r_1)\, d \log_e N_1/dt}{N_2}$$

and found it equal to 3.15; he concluded, therefore, that the two yeasts could not have been regulated by a common factor, and suggested that *Schizosaccharomyces* was somehow poisoning its competitor.

Gause's formula is not easy to use, for it depends so utterly on the accuracy of the equations used to describe the competition. His value of α applies to the period of rapid growth, when $d \log N_1/dt$ is easiest to measure, while ours applies near equilibrium.

One cannot be too careful in drawing conclusions from data of this kind.

Bibliographical Notes

For ways in which even isolated islands are colonized, see G. G. Simpson's lectures, "Geography and Evolution," reprinted in his book, *The Geography of Evolution*, Chilton, 1965. The resistance of different communities to invasion is discussed by C. Elton in *The Ecology of Invasions by Animals and Plants*, Methuen, 1958. Indeed, resistance to outside invaders may be the best measure of a community's adaptedness (evolutionary maturity). Thus the flora of our eastern seaboard has preserved its distinctive character despite three hundred years of European settlement, while a much briefer presence in the Hawaiian lowlands has nearly obliterated the native biota. Some say the native biota disappears more rapidly from settled areas of the tropics than from those in Europe or North America: does a tropical climate somehow facilitate invaders?

Perhaps the best introduction to the plant world is E. J. H. Corner's *The Life of Plants* (reprinted by Mentor, 1968). Corner sees the tropical forest as the norm from which temperate plants deviated, which helps to correct the natural tendency to treat one's homeland as a random sample of the world. He is primarily concerned with the meaning of plant forms, teaching us, in effect, how to understand what we see. MacArthur and Connell's book, *The Biology of Populations*, Wiley, 1966, is only peripherally concerned with plant ecology, but it develops a sense for interpreting landscapes. A detailed description of tropical forest is given in P. W. Richards, *The Tropical Rain Forest*, Cambridge University Press, 1952; parts of the book bristle with unfamiliar plant names, but other parts are of very general interest.

The meanings of tree shape are discussed in H. Horn, *The Adaptive Geometry of Trees*, Princeton University Press, 1971.

John Cairns, Jr. (the limnologist, not the electron microscopist) discusses protozoan interactions in his paper, "Probable Existence of Synergistic Interactions among Different Species of Protozoa," pp. 103-108 of *Revista de Biología*, vol. 6, 1967.

Insects are another, and very important, source of niche differentiation among plants: they have forced plants to evolve an incredible array of defenses to keep from being eaten. See P. R. Ehrlich and P. Raven, "Butterflies and Plants: a Study in Coevolution," pp. 586-608 of *Evolution*, vol. 18, 1965. Most of these defenses are chemical, but some acacias replace the chemicals by private ant colonies which keep off entangling vines as well as insects: see D. Janzen, "Coevolution of Mutualism between Ants and Acacias in Central America," pp. 249-275 of *Evolution*, vol. 20, 1966. How much do these defenses cost? It is hard to measure the energetic cost of a chemical defense, but presumably the cost of maintaining an ant colony is comparable, and more easily measured. Insects also increase the cost of reproduction: see D. Janzen, "Seed-eaters vs. Seed Size, Number, Toxicity and Dispersal," pp. 1-29 in *Evolution*, vol. 23, 1969.

Many chemical interactions are discussed in E. Sondheimer and J. Simeone, eds., *Chemical Ecology*, Academic Press, 1970, including chemical interactions between plants, and their defenses against insects. A fascinating chapter by Carroll Williams shows that some plants control their insect pests by making analogues of their growth hormones, just as we use *plant* hormones to kill weeds.

Diversity increases toward the tropics for animals and plants alike: see A. G. Fischer, "Latitudinal Variation in Organic Diversity," pp. 64-81 of *Evolution*, vol. 14, 1960, and R. H. MacArthur, "Patterns of Species Diversity," pp. 510-533 of *Biological Reviews*, vol. 40, 1965.

Rosenzweig describes his productivity researches in his paper, "Net Primary Productivity of Terrestrial Communities: Prediction from Climatological Data," pp. 67-74 of the *American Naturalist*, vol. 102, 1968.

CHAPTER 10

Animal Communities

In a sense, animals are superfluous. Plants could easily get on without them. Animals become important only when, like ourselves or swarming locusts, they burst their appointed bounds and get out of control. Yet, without insects, there would be no flowers; without birds and squirrels, a woodland would be a silent, lifeless place. Because animals so enliven a landscape, the best of the old Dutch flower painters peopled his pictures with butterflies and ants, caterpillars and snails. In his spirit, we shall now enquire into the ecology of animals.

How are animal niches distinguished? Most animal populations are governed by shortage of prey or abundance of predators: species differ, for the most part, by what eats them, what they eat, or where they feed. Snails and sea urchins sharing the same algae coexist because so few animals eat them both: they are limited by different predators. A rabbit-eating bobcat is too small to attack safely a full-grown buck and too large to live off grasshoppers: some animals coexist by eating different sizes of prey. Bobcats, which stalk their prey and catch them by brief surprise, frequently eat porcupines, while coyotes, which normally run their prey to exhaustion, cannot cope with such a spiny animal: coyotes and bobcats coexist partly because their hunting styles predispose them toward different prey. On the other hand, it does not pay wood warblers to distinguish overmuch between the kinds of insects they find: they choose their food by choosing a place to feed.

An animal's structure reflects its diet, and often its means of escaping predators as well. Species specializing on different prey or attracting different predators will show obvious skeletal differences, detectable even in fossils: one would not confuse a fossil wolf with a coyote or a mountain lion, or a fossil snail with a sea urchin. On the other hand, species differentiated by where they feed may be quite alike in structure. The skeletons of a spruce forest's warblers are quite similar: one tells wood warblers apart by their color, their flight pattern, and where they feed (how far off the ground, and how far into the foliage). "*Paramecium aurelia*" was long considered one species, although in fact it comprises sixteen populations which do not interbreed: these species seem to favor different habitats, and show appropriate differences in temperature tolerance and multiplication rate.

Consider a simple community with two kinds of prey. Many species will invade: will a generalist predator, unable to distinguish between the two prey, eventually oust the others, or will each prey attract its own specialist? Obviously, if the two prey are so unlike that no animal could cope effectively with both, a pair of specialists will starve out a generalist, while two very similar prey will probably attract a generalist. How can we predict the outcome in less clear-cut situations?

Consider an invading species x: suppose it can subsist either on a population R_{1x} of the first prey or R_{2x} of the second. If the logarithmic growth rate of species x increases in proportion to the abundance of each prey, then

$$d \log N_x/dt = e_x(n_1/R_{1x} + n_2/R_{2x} - 1),$$

where N_x is this invader's population, n_1 and n_2 the abundances of the two prey, and e_x a constant measuring the speed at which the population adjusts to its food supply. The prey levels (n_1, n_2) just maintain this invader if $n_1/R_{1x} + n_2/R_{2x} = 1$. If we graph this equation on a plane with axes n_1 and n_2 (Fig. 10-1), we obtain a straight line connecting the points $(R_{1x}, 0)$ and $(0, R_{2x})$, the invader's *isocline*. Points lying inside

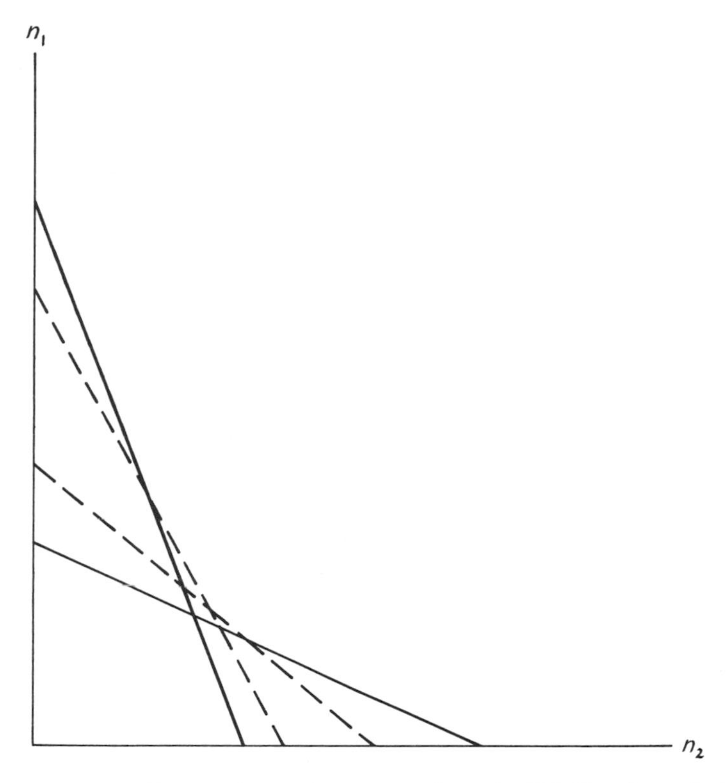

Fig. 10-1. Isoclines of species x and y (solid lines) and their competitors (broken lines). The axes n_1, n_2 of the graph represent abundances of the two prey. Notice that no more than two isoclines intersect at any one point.

this isocline represent prey levels forcing species x to decrease; points outside permit its increase. Now graph the isocline for a second invader, y. The intersection of the two isoclines represents prey levels maintaining both x and y in equilibrium. If x and y are the only two predators colonizing the community, and if there were sufficient prey to support them both, x and y would eat the prey down to this level of joint equilibrium. If one graphs isoclines representing any finite number of invaders, one finds no more than two isoclines intersecting at any one point. No prey level maintains more than two of the invaders in equilibrium: the community can only support two kinds of predators.

Call the invader with the lowest value of R_{1x} the specialist on the first prey, that with the lowest R_{2x} the specialist on the second, and that for which $R_{1x} = R_{2x}$ the generalist. Graph the isoclines of these three species. If the specialist on one prey cannot eat the other, the specialist isoclines will be $n_1 = R_{1x}$ and $n_2 = R_{2x}$. Usually, however, the specialist on one can also live on the other, so its isocline crosses both axes. In a world of mice and rabbits, lynxes specializing on rabbits could, if necessary, live on mice. Because lynxes control mice less effectively than house cats, which specialize on them, the lynx isocline intersects the mouse population axis farther out than does the house cat isocline. A jack of all trades is master of none: a cat larger than a house cat but smaller than a lynx would control mice more effectively than the lynx, but less so than the house cat: it would be better than a house cat, but worse than a lynx, at controlling rabbits. The generalist isocline, therefore, intersects each axis between the specialist isoclines.

If the specialist isoclines cross each other inside that of the generalist, the two specialists can starve out the generalist: if the specialist isoclines intersect inside all the others, the specialists cannot be displaced by any other invader. If the specialist isoclines cross outside that of the generalist (Fig. 10-2) the generalist will starve out at least one specialist, but the community will usually be unstable: an invader can replace one of the two predators present if it has food requirements intermediate between them. After a few such replacements, the predators in

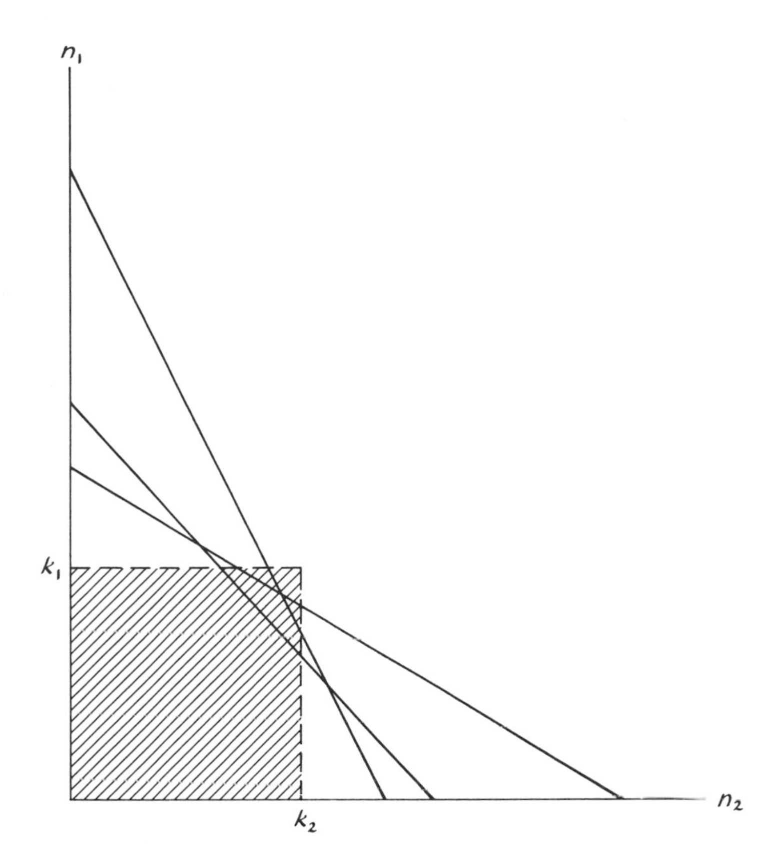

Fig. 10-2. Predator isoclines (generalist favored). The solid lines are isoclines of the generalist and two specialists, and the shaded area represents possible prey abundances: the maximum abundances of the two prey are assumed to be K_1 and K_2. Notice that the food supply does not permit a specialist to coexist with the generalist.

the community will be very similar. If a time comes when one prey finds itself less tolerant than the other of both predators present, it will die out: animals sharing the same habitat and food must attract different predators to coexist. A generalist will survive, however, if neither prey alone could support a specialist (see Fig. 10-2). A generalist can also survive by "switch-feeding": the snail *Thais lapillus* often feeds alternately on barnacles and mussels, eating one prey until it becomes rare, and then focusing on the other. This may enable the snail to manipulate prey levels to the disadvantage of competitors. Switch-feeders, however, must distinguish different kinds of prey, thereby sacrificing an advantage of generalizing: *Thais* apparently feeds thus because it must "adapt" its gut to either barnacles or mussels, but it cannot rely on barnacles alone because barnacle populations fluctuate so violently.

When one kind of animal eats another, the populations of both may fluctuate rather violently: predator-prey relationships are not inherently stable. How can animal populations be stabilized? To answer, we must know how one population is affected by fluctuations in others. Consider a community of n species, which interact only be eating each other. The equation we use to relate the population change of the ith species in the community to the abundances of its predators and prey is

$$dN_i/dt = e_i N_i + a_{i1} N_i N_1 + a_{i2} N_i N_2 + \ldots + a_{in} N_i N_n$$

$$= e_i N_i + \sum_{s=1}^{n} a_{is} N_i N_s.$$

N_i is the population of species i and N_1, N_2, . . . N_n are the populations of the other species in the community. We measure these populations by weight, as we are concerned with their food value. The constant e_i is the logarithmic growth rate of species i when no other animals are present: e_i is positive if species i is herbivorous and negative otherwise. a_{is} is a constant measuring the effect of species s on the population growth of species i: we assume species s affects species i in proportion to the number of meetings between their members, which is proportional in turn to the product of their population sizes. a_{is} is positive if species i eats species s, negative if species s eats species i, and zero if neither eats the other. (If a_{is} is positive, then a_{si} is negative, and vice versa. Why?) If species i eats species s, then $-a_{si}N_i N_s$ is the rate at which species s supplies food to species i. These coefficients a_{is}, which play such a large role in our equations, describe the community's foodweb. We may diagram the foodweb, plotting the species of the community as points, with arrows leading from a species to each of its predators to indicate who eats whom. Plot the herbivores (the species with positive e_i) as a row at the bottom; plot the species feeding mainly on the herbivores, the *primary carnivores*, as the row next above; then the secondary carnivores; etc. If species i eats species s, write the numerical value of $-a_{si}N_i N_s$ over the appropriate arrow to indicate the rate of feeding. Although it is hard to measure feeding rates precisely, one may outline a foodweb from field observations or stomach analyses (Fig. 10-3).

What do foodwebs look like? Farmers tell us it takes ten pounds of feed to make a pound of beef; few species supply their predators more than a tenth of the food received from their prey. If an eagle eats a snake which ate a bird which ate a spider which ate a leafhopper, it receives only a ten-thousandth the food value of the leafhopper. Thus the food chain linking herbivore to terminal carnivore (a carnivore which nobody eats) rarely has more than five links. Moreover, most members of a food chain are larger than their prey: terminal carnivores are rarely abundant animals. When hard times come, they must shorten their food chains to avoid starving: in overpopulated countries, man becomes a vegetarian. If the herbivores eat only plants (never each other), and the primary carnivores eat only herbivores, etc., the community is said to be stratified into trophic

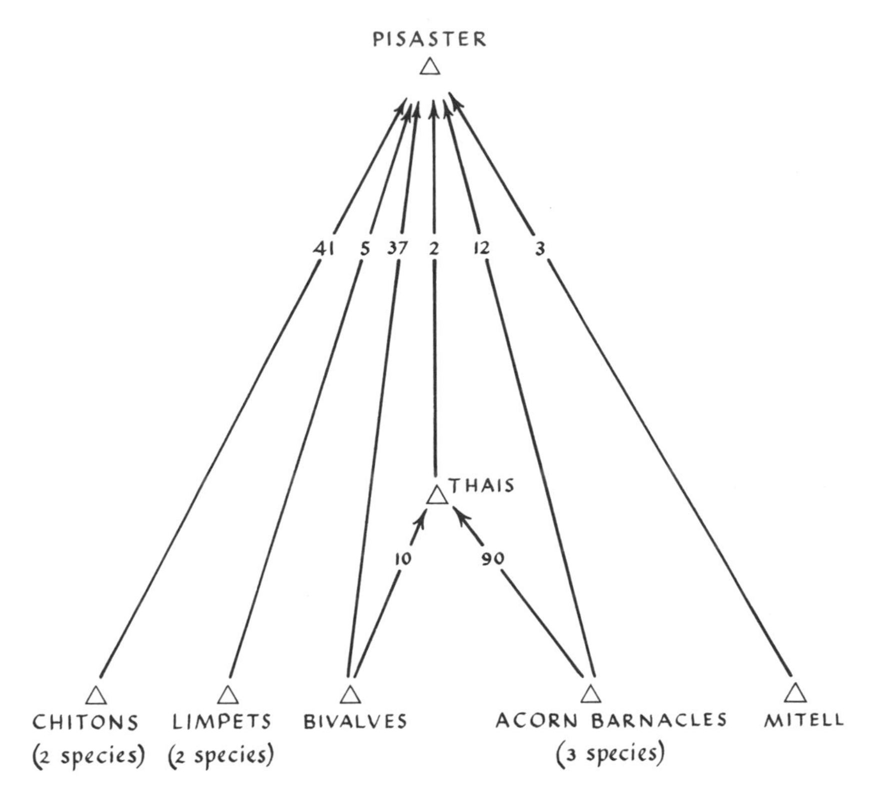

Fig. 10-3. A portion of a foodweb: food chains leading to the terminal carnivore *Pisaster* (a large starfish). Lines leading up to a species indicate what it eats: the numbers indicate the percentage (in calories) each prey contributes to the predator's diet. This figure diagrams an intertidal community from the Pacific coast of Washington. (After Fig. 1 of R. T. Paine, "Food Web Complexity and Species Diversity," the *American Naturalist*, p. 67, vol. 100, 1966. Used with permission of the author and of the University of Chicago Press.)

levels: a textbook example (which happens to be wrong) is a pond where large fish eat only small fish, which eat herbivorous insects and snails. Herbivores usually form a clear-cut trophic level: rabbits and deer do not eat other animals. A weasel, however, would as happily eat another weasel as a baby rabbit: carnivores do not form neat trophic levels. No community is neatly stratified into herbivores, primary carnivores, and secondary carnivores: the primary carnivores could not simultaneously maintain their prey and predators in equilibrium*. If there are secondary carnivores, they eat herbivores as well.

What governs the stability of a species? A species is stable if its population rarely deviates far from its mean, and unstable if its population frequently "crashes" or "explodes." Instability endangers survival: a crashing population may go extinct, and an exploding one may upset an essential balance. Intuitively, we would expect larger populations to be less susceptible to chance fluctuations in numbers, species with more diverse diets to be less susceptible to crashes or explosions in one of their prey, and populations of longer-lived individuals to change more slowly, and crash or explode less often. Since all species tolerate much the same risks of extinction, those living in unstable environments require more effective stabilizing mechanisms.

---

* Consider a community composed of an herbivore, a primary carnivore, and a secondary carnivore: let their populations be N_1, N_2, and N_3, respectively. The equations of population change will be

$$dN_1/dt = e_1N_1 + a_{12}N_1N_2,$$
$$dN_2/dt = e_2N_2 + a_{21}N_1N_2 + a_{23}N_2N_3,$$
$$dN_3/dt = e_3N_3 + a_{32}N_2N_3.$$

e_2, e_3, a_{12}, and a_{23} are negative constants; e_1, a_{21}, and a_{32} are positive. If the herbivores are in equilibrium, $dN_1/dt = 0$ and $N_2 = -e_1/a_{12}$. If the secondary carnivores are in equilibrium, $dN_3/dt = 0$ and $N_2 = -e_3/a_{32}$. As N_2 cannot assume two values at once, the herbivores and secondary carnivores cannot be simultaneously in equilibrium. Rosenzweig shows that two species of weasel feeding on a common, partially space-limited prey can coexist if the larger weasel also eats the smaller: apparently, a secondary carnivore can stabilize his niche by eating the herbivore as well as the primary carnivore.

The changeable, unpredictable Arctic is the home of species with large populations and diverse diets, while the stabler and more productive tropics harbor multitudes of rare specialists. A terminal carnivore compensates for its low population by an unusually diverse diet. A species whose food supply is slowly dwindling may stabilize its population by living longer. The Paleozoic and Mesozoic eras seem both to have ended with a slow decline in the supply of marine plankton. The filter-feeders which lived on it could not widen their diet: reef brachiopods of the Permian and rudistid clams of the Cretaceous responded by slowing their pace of life, both evolving long-lived giant forms shortly before dying out altogether, apparently for lack of food.

Our equations are far too crude to calculate the exact course of a population's changes: how can we predict stability? The early students of gases were faced with crude equations describing gas molecules as minute billiard balls. They focused on averages, hoping that averages over large numbers of molecules would somehow smooth out the errors in their assumptions. This worked: they derived a relation between the average energy of gas molecules and how often these molecules collided with the walls of their container which, properly interpreted, yielded the perfect gas law. The precise derivation of gas laws from atomic physics is the province of statistical mechanics, whose techniques also permit us to calculate the stability of populations in diverse communities (we depend on averages over a large number of species to smooth out errors in our assumptions). To apply these methods, we must measure populations in herbivore equivalents: if ten pounds of deer are required to make a pound of puma, we value a pound of puma as ten herbivore-pounds. We value a pound of secondary carnivore as 100 herbivore-pounds, etc. We cannot value a carnivore eating both weasels and squirrels: these units only work for communities stratified into trophic levels. We shall henceforth assume that this is the case, hoping our error will not mislead us.

Let us agree that a population crashes when it falls to a fifth its average abundance and explodes when it exceeds five times its average. The instability, S, of a species is the frequency of crashes or explosions in its population. If populations rarely deviate far from their average levels, the instability S_i of the ith species in a community of n approximates

$$(1/2\pi)\sqrt{\sum_{s=1}^{n} a_{is}{}^2 \overline{N_i N_s}} \quad \exp\left[-\overline{N}_i (\ln 5)^2 / 2\theta\right],$$

where all species s which are predators or prey of species i contribute to this sum. $\overline{N}_i$ and $\overline{N}_s$ are the average populations of species i and s, e = 2.718.. is the base of the natural logarithms, ln 5 is $\log_e 5$, and exp $[x]$ is e^x. θ is a constant measuring the variability of the physical environment: it is larger for more disturbed environments.* This expression for S_i is the more accurate the lower θ.

Suppose first that species i is a terminal carnivore which exerts no preferences among its prey. a_{is} will then be the same for all species s it eats (see Appendix 2). Its average feeding rate, $\overline{F}_i$, will be

$$\sum_{s=1}^{n} a_{is}\overline{N_i N_s} = \sum_{s=1}^{n} a_{is}\overline{N}_i\overline{N}_s \doteq a_{is}\overline{N}_i\overline{N}_p \,.$$

$\overline{N}_p$ is the total population of the species eaten by species i. (We can replace $\overline{N_i N_s}$ by $\overline{N}_i\overline{N}_s$ because our assumptions imply that the populations of different species are statistically inde-

---

* a) If we say the population crashes when it falls to $1/A$ its average level, and explodes when it exceeds A times its average, then S_i is

$$(1/2\pi)\sqrt{\sum_{s=1}^{n} a_{is}{}^2\overline{N}_i\,\overline{N}_s} \exp\left[-\overline{N}_i\,(\ln A)^2\,/2\theta\right].$$

b) θ is the average, for all species in the community. of $\overline{(N_i - \overline{N}_i)^2}\,/\overline{N}_i$. $\overline{(N_i - \overline{N}_i)^2}$ is the mean square deviation of the population of species i from its average.

pendent.) Expressing a_{is} as $\overline{F}_i/\overline{N}_i\overline{N}_p$, the instability S_i of species i is

$$1) \quad (1/2\pi)\sqrt{\overline{F}_i^2/\overline{N}_i\overline{N}_p}\ e^{-k\overline{N}_i},$$

where $k = (\ln 5)^2/2\theta$. Stability is greater the larger the population $\overline{N}_i$, and the more diverse its diet (the larger $\overline{N}_p$). A population of N herbivore equivalents eating F units of food a year renews itself every N/F years: $\overline{N}_i/\overline{F}_i$ is thus the average lifetime of species i. Calling this lifetime $\overline{L}_i$, S_i may be expressed as

$$(1/2\pi)\sqrt{\overline{F}_i/\overline{N}_p\overline{L}_i}\ e^{-k\overline{N}_i}.$$

A filter-feeder eating all it catches cannot change $\overline{F}_i$ or $\overline{N}_p$: the only way to stabilize its population is to increase $\overline{L}_i$.

If species i is not a terminal carnivore, we must consider the effect of its predators. If a_{ir} is the same for all species r preying on species i, then the instability of species i is

$$(1/2\pi)\sqrt{\overline{F}_i/\overline{N}_i\overline{N}_p + \overline{F}_i'^2/\overline{N}_i\overline{N}_{Pr}}\ e^{-k\overline{N}_i},$$

where $\overline{N}_{Pr}$ is the total population of its predators, and $\overline{F}_i'$ the average rate at which they feed on species i. A population is stabler if it has no predators at all, but if it must be eaten, it is better off preyed on lightly by many species than heavily by a few.

How do the characteristics of the community as a whole, its productivity, its standing crop, the connectedness of its foodweb, etc., affect the stability of its component species? To answer, we must define our terms. The community's standing crop or "biomass," which we call B, is the sum of the populations of all species present: in symbols,

$$B = \sum_{i=1}^{n} \overline{N}_i.$$

B is the weight of the herbivores, plus ten times the weight of the carnivores, etc.: herbivore equivalents emphasize the higher trophic levels, which are ecologically more important than their weights suggest. Community productivity, P, is the rate at which animal matter is formed. If most animals die from predation, productivity is nearly equal to the rate at which animal matter is eaten. In symbols, P is

$$\sum_{i,s} -a_{is} N_i N_s \,.$$

The sum includes every pair of species i and s such that i eats s. The interconnectedness, I, of the foodweb is the average number of links per species (the number of links in the web, divided by the number of species).

Suppose, for simplicity, that the average population of each species is B/n, that each species is directly linked to $2I$ others (if there are I links per species, the "average" species has I kinds of prey and I kinds of predator), and that all coefficients a_{is} differing from 0 have the same absolute value, which we call a. Community productivity is the number of links in the foodweb (the number of species in the community times the number of predators the average species has), times the amount of food flowing along each link: in symbols,

$$P = nI(aB^2/n^2).$$

Expressing a in terms of the other variables, we find

$$a = nP/B^2 I.$$

The instability of each species in the community is then

$$(1/2\pi)\sqrt{\textstyle\sum_s a^2 \bar{N}_i \bar{N}_s}\;\; e^{-\bar{N}_i (\ln 5)^2/2\theta}$$

$$= (1/\pi)(P/B)\sqrt{1/2I}\;\; e^{-B(\ln 5)^2/2n\theta}.$$

Communities with a large standing crop and a low productivity-biomass ratio should be particularly diverse: specialists are less likely to die out under these conditions, so more species can partition the available resources.

Appendix 1

A community containing the two carnivores s and s'' can be invaded by a third, s', with intermediate food requirements, only if the isoclines of s and s'' intersect outside that of s'. What is the algebraic condition for this? Suppose the isoclines of the three species are (Fig. 10-4)

$$\frac{x}{R_1(s)} + \frac{y}{R_2(s)} = 1, \quad \frac{x}{R_1(s')} + \frac{y}{R_2(s')} = 1,$$

$$\frac{x}{R_1(s'')} + \frac{y}{R_2(s'')} = 1,$$

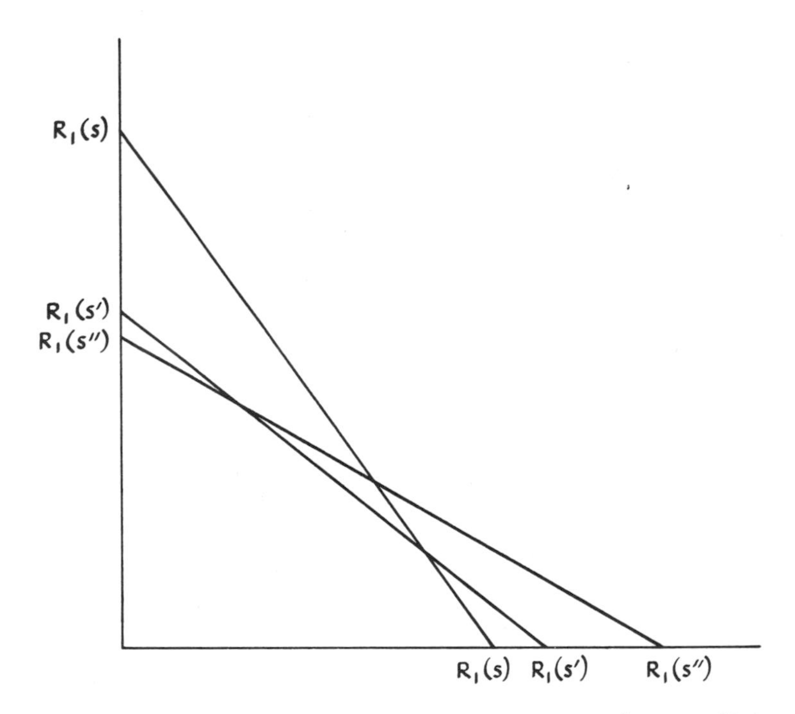

Fig. 10-4. Isoclines of the three predators, s, s', and s'', discussed in Appendix 1.

where x is the abundance of the first prey and y that of the second, and where $R_1(s) > R_1(s') > R_1(s'')$, $R_2(s'') > R_2(s') > R_2(s)$. If s' can increase on the food supply x', y' maintaining s and s'' in equilibrium, then

$$2) \quad \frac{x'}{R_1(s')} + \frac{y'}{R_2(s')} > 1$$

for that point x', y' where the isoclines of s and s'' intersect. This point of intersection is the solution of the simultaneous equations

$$\frac{x'}{R_1(s)} + \frac{y'}{R_2(s)} = 1, \quad \frac{x'}{R_1(s'')} + \frac{y'}{R_2(s'')} = 1,$$

to wit:

$$x' = \frac{\dfrac{1}{R_2(s)} - \dfrac{1}{R_2(s'')}}{\left(\dfrac{1}{R_1(s'')} - \dfrac{1}{R_1(s)}\right)\dfrac{1}{R_2(s)} + \left(\dfrac{1}{R_2(s)} - \dfrac{1}{R_2(s'')}\right)\dfrac{1}{R_1(s)}},$$

$$y' = \frac{\dfrac{1}{R_1(s'')} - \dfrac{1}{R_1(s)}}{\left(\dfrac{1}{R_1(s'')} - \dfrac{1}{R_1(s)}\right)\dfrac{1}{R_2(s)} + \left(\dfrac{1}{R_2(s)} - \dfrac{1}{R_2(s'')}\right)\dfrac{1}{R_1(s)}}.$$

Substituting for x' and y' in equation 1 and rearranging, we obtain

$$\frac{1}{R_1(s')}\left[\frac{1}{R_2(s)} - \frac{1}{R_2(s'')}\right] + \left[\frac{1}{R_1(s'')} - \frac{1}{R_1(s)}\right]\frac{1}{R_2(s')}$$

$$> \frac{1}{R_1(s)}\left[\frac{1}{R_2(s)} - \frac{1}{R_2(s'')}\right] + \left[\frac{1}{R_1(s'')} - \frac{1}{R_1(s)}\right]\frac{1}{R_2(s)}.$$

We may rewrite this as the inequality

$$\text{(a)}\quad \left[\frac{1}{R_1(s')} - \frac{1}{R_1(s)}\right]\left[\frac{1}{R_2(s)} - \frac{1}{R_2(s'')}\right]$$

$$> \left[\frac{1}{R_2(s)} - \frac{1}{R_2(s')}\right]\left[\frac{1}{R_1(s'')} - \frac{1}{R_1(s)}\right].$$

The condition of continued invasibility is that the inequalities $R_1(s) > R_1(s') > R_1(s'')$ and $R_2(s'') > R_2(s') > R_2(s)$ always imply the inequality (a).

Appendix 2

A terminal carnivore stabilizes its population by exerting no preference among the prey it feeds on. Let

$$a_{is} = \frac{1}{N_p} \Sigma'_s\, a_{is} N_s\,,$$

where N_p is the total population of the prey of species i and the sum Σ'_s includes all species s eaten by species i. To show that $\Sigma'_s a^2_{is} N_i N_s$ is minimum if all the a_{is} are equal to $\bar{a}_{is}$, we prove that

$$\Sigma'_s a^2_{is}\, N_i N_s - \Sigma'_s\, \bar{a}^2_{is}\, N_i N_s = \Sigma'_s\, (a_{is} - \bar{a}_{is})^2\, N_i N_s \geqslant 0:$$

this last sum vanishes only if $a_{is} = \bar{a}_{is}$ for all species s eaten by species i.

$$\Sigma'_s\, a_{is} N_i N_s - \Sigma'_s \bar{a}^2_{is} N_i N_s = \Sigma'_s\, (a^2_{is} - \bar{a}^2_{is})\, N_i N_s =$$

$$\Sigma'_s\, (a_{is} - \bar{a}_{is})(a_{is} + \bar{a}_{is}) N_i N_s =$$

$$\Sigma'_s\, (a_{is} - \bar{a}_{is})^2\, N_i N_s + 2\bar{a}_{is} N_i\, \Sigma'_s\, (a_{is} - \bar{a}_{is}) N_s\,.$$

Since

$$\Sigma'_s a_{is} N_s = \Sigma'_s \bar{a}_{is} N_s = \bar{a}_{is} N_p \text{ ,}$$

the sum multiplying $2\bar{a}_{is}N_i$ must be zero, establishing the assertion.

Problems

1. Calculate the connectivity of the web diagrammed in Fig. 10-3 (consider each node of the web a species).

Solution: In Fig. 10-3, chitons, limpets, *Mitella*, and *Thais* are each eaten by one species, bivalves and acorn barnacles are eaten by two apiece, while nothing preys on *Pisaster*. Thus four species "initiate" a link apiece, two initiate two, and one initiates none at all: the web's connectivity (the average number of links per species) is thus 8/7, or 1.14.

2. Consider a terminal carnivore with a population of N herbivore-equivalents, eating F units of food a year, feeding on prey whose populations total N_p. This population can stabilize either by diversifying its diet, preying lightly on many species rather than heavily on a few (increasing N_p but not F), or by increasing its population, N. N may be increased either by increasing F (assume that in this case N/F is constant) or by increasing N/F (lengthening life).

   a) Is diversifying diet relatively more effective in stable or unstable environments?

   b) Are there any circumstances where a population can stabilize by preying more heavily on existing prey?

Solution:

a) Let this carnivore's instability be S. S is $(1/2\pi)\sqrt{F^2/NN_p}$

$\exp[-(\ln 5)^2 N/2\theta]$. To compare the effects of different changes, take the differential of $\log S$.

$$\log S = -\log 2\pi + \log F - (1/2)(\log N + \log N_p) - (\ln 5)^2 N/2\theta,$$

$$d \log S = d \log F - (1/2)(d \log N + d \log N_p) - (\ln 5)^2 dN/2\theta.$$

If N is increased by increasing F in such a way that F^2/NN_p is constant, $d \log S = -(\ln 5)^2 dN/2\theta$: the stability increase is less the larger θ (the more unstable the physical environment); $d \log S$ is negative because it refers, strictly speaking, to a decrease in instability. If N increases but not F, $d \log S$ is $-(1/2)d \log N - (\ln 5)^2 dN/2\theta$: the stability increase is less for large θ, but θ's effect is less than before. If diet is diversified without increasing F, $d \log S = -d \log N_p$, which is the same whatever the stability of the physical environment. Diversifying diet is *relatively* more effective in unstable environments.

b) If F increases without increasing N_p, $d \log S$ is $d \log F - (1/2)d \log N - (\ln 5)^2 dN/2\theta$. Presumably N increases in proportion to F, so $d \log F = d \log N$, and we may express $d \log S$ as $(1/2)d \log N - (\ln 5)^2 dN/2\theta$. This strategy increases stability if

$$(\ln 5)^2 dN/2\theta > (1/2)d \log N.$$

Remembering that $d \log N = dN/N$, we may rephrase this inequality as

$$(\ln 5)^2/2\theta > 1/2N.$$

This strategy is most likely to work for large populations in stable environments.

Notice that if $\theta = \overline{(N - \overline{N})^2}/\overline{N}$, our condition becomes $\overline{(N - \overline{N})^2}/(\ln 5)^2 < N^2$.

Bibliographical Notes

R. MacArthur and R. Levins analyze the distinction between generalists and specialists in a very difficult paper, "Competition, Habitat Selection and Character Displacement in a Patchy Environment," pp. 1207-1210 of the *Proceedings of the National Academy of Sciences*, vol. 51, 1964; the distinction is explored further in R. MacArthur, "Patterns of Species Diversity," pp. 510-533 of *Biological Reviews*, vol. 40, 1965. Structural consequences of the distinction are discussed in G. E. Hutchinson, "The Sensory Aspects of Taxonomy, Pleiotropism, and the Kinds of Manifest Evolution," pp. 533-540 of the *American Naturalist*, vol. 100, 1966. T. Sonneborn discusses the syngens of *Paramecium* in "Breeding Systems, Reproductive Methods, and Species Problems in Protozoa," pp. 155-324 of E. Mayr, ed., *The Species Problem*, American Association for the Advancement of Science Publication No. 50, 1951. Niche differentiation among carnivorous mammals is discussed in Rosenzweig, "Community Structure in Sympatric Carnivora," pp. 602-612 of the *Journal of Mammalogy*, vol. 47, 1966, and that among warblers in R. MacArthur, "Population Ecology of some Warblers of Northeastern Coniferous Forests," pp. 599-619 of *Ecology*, vol. 39, 1958. The switch-behavior of *Thais* is described in E. Fischer-Piette, "Histoire d'une Moulière," pp. 152-177 in *Bulletin Biologique*, vol. 69, 1935.

Our catalogue of the modes of niche differentiation among animals is hardly complete. R. T. Paine tells me that anemones of the Washington coast flourish when large starfish are working nearby, for many scraps float their way. This is rather like the relation between jackals and lions. These are essentially "third-order interactions," involving prey, predator, and scavenger all at once, and thus do not fit easily into our scheme of Volterra equations.

Our mathematics exerts other subtle limitations on our thinking. We describe a population by its number of individuals, neglecting to distinguish their ages: what are we to make of the Pacific mussels, whose numbers are recruited primarily from those few individuals who grow lower than usual in the intertidal, where starfish are common, and who happened to escape predation long enough to grow too big to eat? A counting or weighing of mussels would emphasize the masses of little ones in the upper intertidal, but the fate of the species may depend on the lucky few which survive lower down.

Lindeman analyzes pond communities in terms of trophic levels in his paper, "The Trophic-Dynamic Aspect of Ecology," pp. 399-418 of *Ecology*, vol. 23, 1942. R. T. Paine discusses some foodwebs in his paper, "Food Web Complexity and Species Diversity," pp. 65-75 of the *American Naturalist*, vol. 100, 1966. Paine emphasizes the importance of predation in maintaining species diversity: if a predator favors the "most efficient competitor," more species will coexist.

Stability is an embarrassing subject. The very word is ambiguous. To mathematicians, a system is stable if it returns to equilibrium when disturbed: the stabler the system, the more quickly equilibrium is restored (Rosenzweig and MacArthur use the word in this sense in their paper, "Graphical Representation and Stability Conditions of Predator-prey Interactions," pp. 209-223 of the *American Naturalist*, vol. 97, 1963; the subject is discussed in most books on differential equations). In this view, the balance between a community's stability and environmental disturbance determines the amplitude of population fluctuations. Biologists are often more vague in their usage: I think their meaning is best captured by calling stability the frequency of a population's crashes or explosions. This is easier to predict and measure than stability as defined by mathematicians. Views on the causes of stability conflict even more. MacArthur argued that the more diverse a community and the more connected its foodweb, the stabler it would be (R. MacArthur, "Fluctuations in Animal Populations and a Measure of Community Stability," pp. 533-536 in *Ecology*, vol. 36, 1955): if any population should explode, its depredations would be spread over many prey, and many predators would be available to control it. MacArthur even suggested that a community's stability be *measured* by the complexity of its foodweb. The rarity of insect outbreaks in complex tropical communities (D. Pimentel, "Species Diversity and Insect Population Outbreaks," pp. 76-86 of the *Annals of the Entomological Society of America*, vol. 54, 1961) supports this view. However, Kenneth Watt (*Ecology and Resource Management*, Prentice-Hall, 1968) finds that species eating only one kind of prey are stabler than those eating many; and R. T. Paine tells me that populations of the sea slug *Navonax* feeding on one kind of prey are stabler than those with diverse diets. The paradox disappears if we remember there is no tendency to "maximize" stability. Species specialize as much as conditions allow: those with only one prey may already be "completely" specialized and unable to take further advantage of their stability.

Our analysis of stability follows E. Leigh, "On the Relation between the Productivity, Biomass, Diversity and Stability of a Community," pp. 777-783 of the *Proceedings of the National Academy of Sciences*, vol. 53, 1965, and "The Ecological Role of Volterra's Equations," pp. 1-61 in M. Gerstenhaber, ed., *Some Mathematical Problems in Biology*, American Mathematical Society, 1968. Unfortunately, both papers are exceedingly mathematical. A verbal argument developed independently along similar lines is J. Connell and E. Orias, "The Ecological Regulation of Species Diversity," pp. 399-414 of the *American Naturalist*, vol. 98, 1964: our discussion can be viewed as a "model" or "realization" of their argument.

Our treatment, however, is very crude. We measure populations as uniform densities, with no way of providing for the effects of a patchy distribution: we are compelled therefore to assume that species are differentiated only by what they eat, never by where they feed. Moreover, we assume that animals do not suit their food preferences to the relative abundances of their prey: our mathematics takes no account of "switch-feeders." Either assumption deprives the analysis of rigor, and may invalidate it entirely.

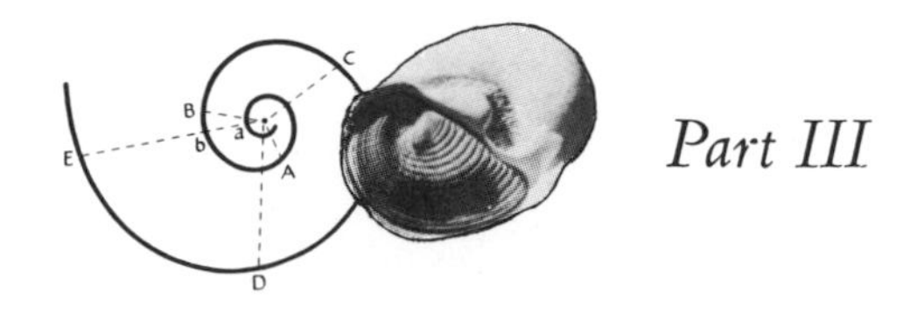

Part III

THE FOSSIL RECORD

Before embarking on evolutionary theory, *it is best to remind ourselves of the nature of the available evidence.*

How does one assemble the story of evolution? How detailed a history can it be? What does it tell us about the predictability of evolution?

CHAPTER 11

Paleoecology

DO SPECIES form (or immigrate) so slowly that no community can ever be fully stocked? Can one explain the hundreds of species of fish in Lake Tanganyika and Lake Nyasa by the fact that these lakes have lasted much longer than most? Has the wealth of brightly colored fresh-water clams of the American southeast evolved because the area has been spared dessication, glaciation, and submergence beneath the sea during the past few million years? Are tropical rain forests so diverse because they are so much older and more widespread than deserts and temperate forests? Are deserts and temperate zones so rare and short-lived that each example must be settled anew by invaders from the rain forest? To answer such questions we must learn about conditions in the past.

We infer past conditions by analogy with the present. Clams live in water, so fossil communities containing clams were aquatic. The scallops of New Jersey live only in the sheltered lagoons behind the barrier beaches, so the blackened scallop shells on the ocean beach must be cast up from fossil lagoon deposits just offshore. The entry of a northern clam into the Mediterranean implies the cooling of that sea: its comings and goings record the temperature changes of the Mediterranean during the ice ages. Moreover, the ratio of the isotopes of oxygen, O^{16} and O^{18}, in shells of these clams reflects the temperature of their water, so the seasonal variation in water temperature can be measured by analyzing the shells. Other analogies tell us of food and feeding habits. Most clams feed by filtering plankton from the water; most limpets graze algae on seaside rocks. A small, round hole in a clam shell means that a snail drilled with its rasping tongue to suck out the juice and soft flesh: an octopus's beak makes a more irregular hole. Some small carnivorous snails have a tooth near the bottom of their outer lip (Fig. 11-1) which apparently anchors the snail when it drills or pries open its prey. The jaws and teeth of a mammal

Fig. 11-1. Opeatostoma, a carnivorous snail. (Photograph by S. Rawson.)

(Fig. 11-2) reveal whether it is herbivorous or carnivorous; if herbivorous, whether it rooted like a pig, grazed like a cow, or browsed on forest leaves like a deer; if carnivorous, the size of its prey and perhaps how it caught them. The lore of natural history supplies seemingly endless analogies to elucidate the past.

Sometimes, answers to our questions lie in subtle quantitative analyses. How prevalent were tropical conditions in the Cretaceous and Eocene? To answer, Bews compared fossil plant communities scattered over Europe and North America with modern forests in Britain and South Africa. For each community, he tabulated the proportions of angiosperm species with simple leaves less than an inch long, between one and three inches, and over three inches; and those with compound leaves. He obtained the following table:

Proportion of species with	*simple leaves > 3 in.*	*simple, 1–3 in.*	*simple, < 1 in.*	*compound leaves*
moist subtropical forest (South Africa)	75%	19%	0%	6%
mesophytic forest (South Africa)	32%	47%	3%	18%
dry parkland (South Africa)	11%	34%	13%	42%
British temperate forest	20%	36%	28%	16%
range for all fossil communities	47-64%	10-29%	2-7%	10-31%
Willcox flora (the fossil community with the least subtropical array of leaves)	47%	17%	5%	31%

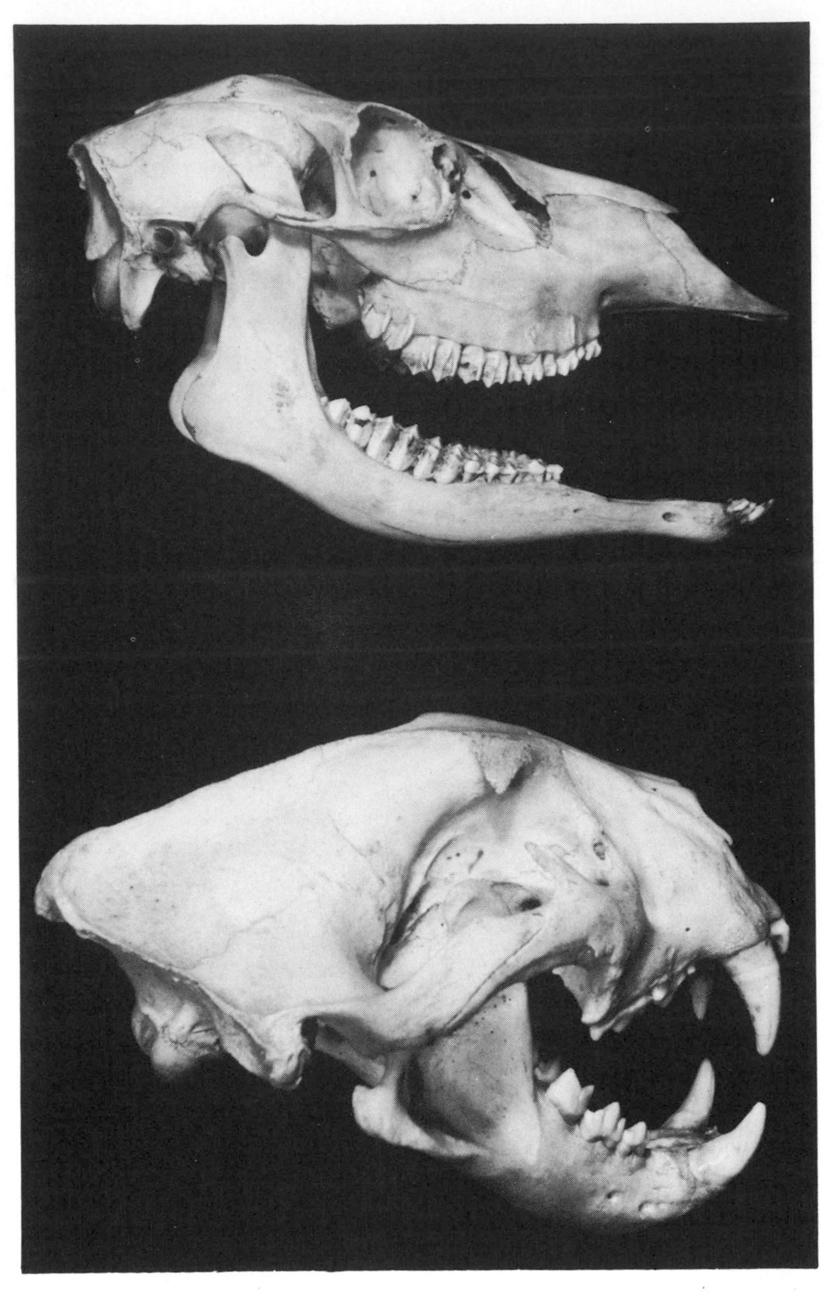

Fig. 11-2. Skulls of mule deer (top) and jaguar (bottom). (Photograph by D. Baird.)

He found, moreover, that in the Willcox flora the proportion of species with entire (as opposed to lobed or toothed) leaves was the same as in modern rain forests, and much greater than in British woodland. Since the leaf forms in the fossil communities resembled those of moist subtropical forest, Bews concluded that warm, moist conditions were widespread in the Eocene. The forest of tropical mountains, however, also bears such leaves, in spite of a disagreeably cold and wet (but frostfree) climate: do such leaves grow in any wet, frostfree region?

We use the present as key to the past, but we justify ourselves only with the claim that the laws of physics and chemistry are unchanging. What of the analogies we customarily use to unlock the past? Life evolves, changing the world: the older a fossil community, the shakier our analogies. If we recognize the species of a fossil community, we know what they did and where they lived (at least if they haven't changed), but not so many million years ago the world was populated by different species. We think of clams as filter-feeders, but when they first evolved they picked up detritus from the bottom. It benefits us progressively less to infer by the *kinds* of animals and plants in a community. Analogies of form are more useful. The forms of organisms reflect their ways of life, and reveal to the knowing eye their habits and living conditions. The syngens* of *Paramecium* (see Chapter 10) warn us, however, that form does not reveal everything about way of life. Empirical analogies of form are dangerous: not knowing the advantages of one leaf shape over another, we cannot be sure what leaf shape means. Moreover, evolution changes the relation of form to way of life: riddles of ancient form are often difficult to read, and some Cambrian fossils are incomprehensible because they are so utterly without modern parallel (Fig. 11-3).

How do mature communities reflect their environments?

---

* Syngens are species indistinguishable in form, which do not interbreed.

The relative abundances of related species reveal the conditions under which they formed: the total number of species reflects the stability and productivity of the community. Such inferences from community ecology require data notoriously difficult to quarry from the fossil record, but the theories involved are so general they apply equally to the Cambrian and the present day.

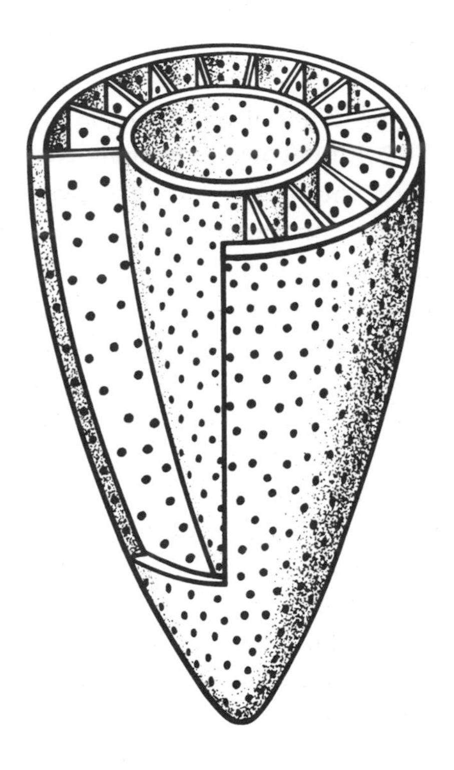

Fig. 11-3. An idealized drawing of an archaeocyathid. No one knows how this animal made its living. (After Fig. 585 from *The Course of Evolution*, by J. M. Weller. Copyright 1969 by McGraw-Hill, Inc. Used with permission of McGraw-Hill Book Company.)

We saw in an earlier chapter how, under stable conditions, the relative abundances of related species follow the MacArthur distribution. Goulden showed that the relative abundances of the species of *Chydoridae* (bottom-feeding cladoceran *Crustacea*, about 1 mm. long) in the sediments of a small Guatemalan pond reflected the pond's equilibrium. On the whole, these chydorids fitted MacArthur's distribution rather well, and the fit improved as the pond aged and their diversity increased (Fig. 11-4). They fitted worst in the oldest sediments, formed while the pond community was still immature. When the pond had nearly achieved a balance, people settled around its borders and burned the surrounding vegetation, thereby fertilizing the pond and temporarily disturbing its balance. This disturbance was reflected by a deviation from MacArthur's distribution: the most common chydorid was too abundant to fit the theory.

Ecology also tells us that the closer a community is to the equator, the greater is its diversity. Stehli used this principle to map the Permian equator. First he verified this principle for different groups of modern animals, tabulating the number of species occurring at each of numerous localities scattered over the earth, and calculating "contours of equal diversity" analogous to contours of equal altitude on a topographic map. The contours of maximum abundance nearly always coincided roughly with the equator (Fig. 11-5), (different methods of calculating the contours, however, often yield discordant equators). Then Stehli applied his procedure to Permian animals, again finding equators concordant with the modern one. His finding shed light on an old dispute. During the Permian there were subtropical forests in Greenland and glaciers in Patagonia, South Africa, and India, implying either that the poles have shifted or that the continents have changed position since. Stehli's finding suggests that this anomaly resulted from continental drift, and that as much land has drifted north as south since the Permian.

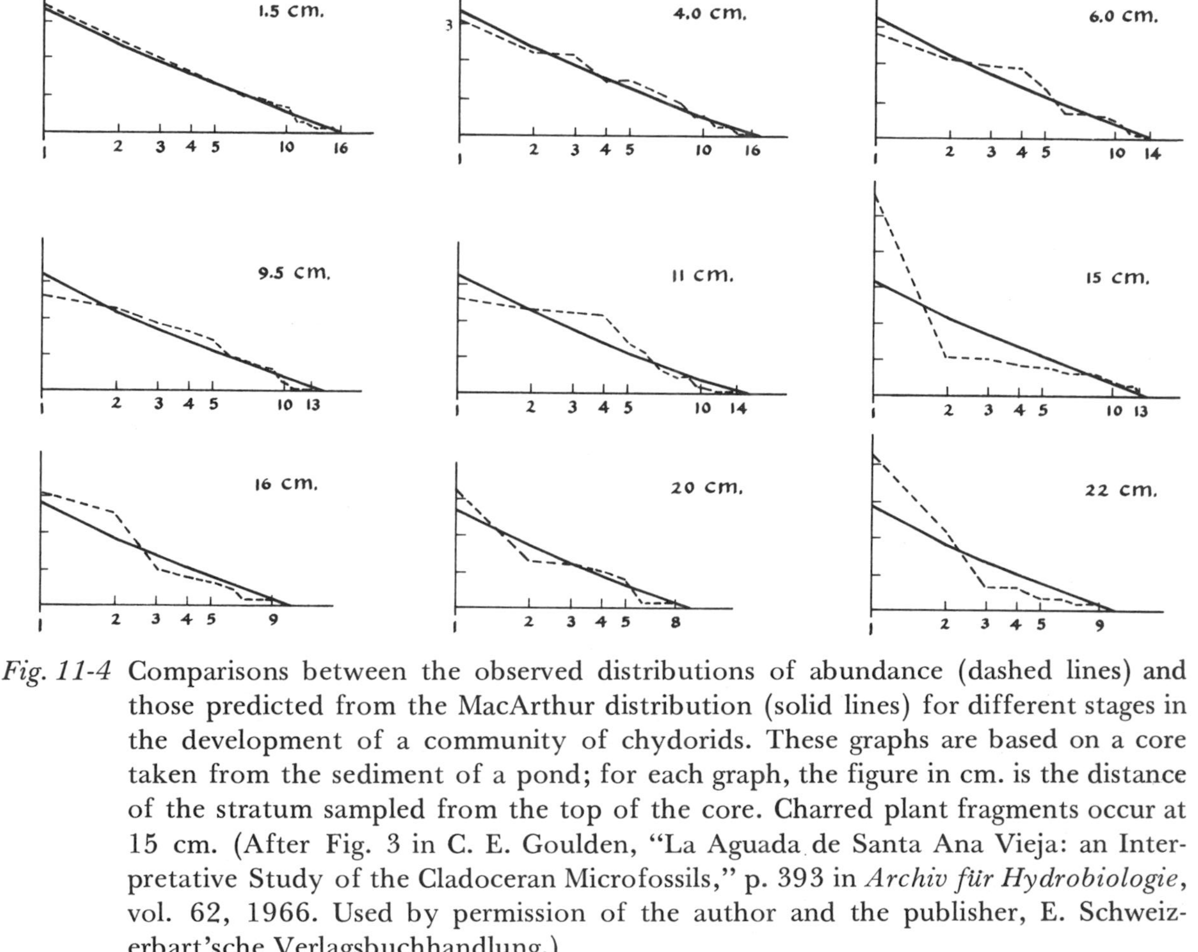

Fig. 11-4 Comparisons between the observed distributions of abundance (dashed lines) and those predicted from the MacArthur distribution (solid lines) for different stages in the development of a community of chydorids. These graphs are based on a core taken from the sediment of a pond; for each graph, the figure in cm. is the distance of the stratum sampled from the top of the core. Charred plant fragments occur at 15 cm. (After Fig. 3 in C. E. Goulden, "La Aguada de Santa Ana Vieja: an Interpretative Study of the Cladoceran Microfossils," p. 393 in *Archiv für Hydrobiologie*, vol. 62, 1966. Used by permission of the author and the publisher, E. Schweizerbart'sche Verlagsbuchhandlung.)

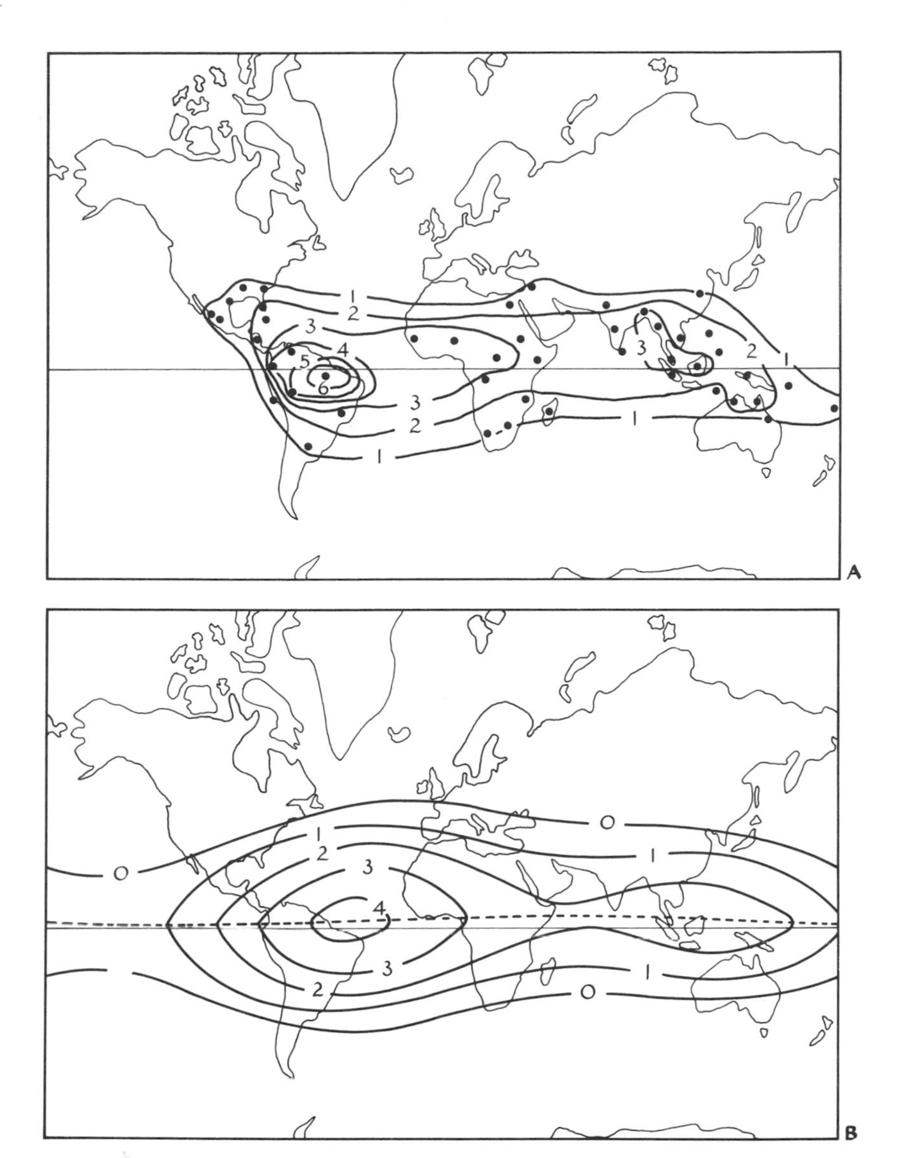

Fig. 11-5. The upper map, A, shows contours of crocodilian diversity; the points are places where species counts were made. The lower map, B, shows the contours of a second-order spherical harmonic fitted to predict the equator from crocodile diversity. (After Figs. 43 and 44 of F. G. Stehli, "Taxonomic Diversity Gradients in Pole Location: The Recent Model," p. 208, in E. T. Drake, *Evolution and Environment*, 1968, with permission of the Yale University Press, New Haven, Connecticut.)

Bibliographical Notes

Paleoecologists are contributing more and more to our understanding of ecology, for they need *meaningful* generalizations to guide their interpretations of the past. Two excellent collections of papers on paleoecology and related topics are Imbrie and Newell, *Approaches to Paleoecology*, Wiley, 1964, and Hedgpeth and Ladd, *Treatise on Marine Ecology and Paleoecology* (2 vols.), Geological Society of America Memoir no. 67, 1957.

In his paper, "Speciation in Ancient Lakes" (pp. 30-60 of *Quarterly Reviews of Biology*, vol. 25, 1950), J. L. Brooks discusses the evolution of the amazing diversity of life in Russia's Lake Baikal, parts of which are over fifty million years old.

Fischer discusses what he learned from the blackened scallops on New Jersey's barrier beaches in "Stratigraphic Record of Transgressing Seas in Light of Sedimentation on the Atlantic Coast of New Jersey," pp. 1656-1666 of the *Bulletin of the American Association of Petroleum Geologists,* vol. 45, 1961. Not everyone is convinced that a knowledge of the species present in a fossil deposit accurately reflects the conditions in which they lived: see Raup and Lawrence, "Paleoecology of Pleistocene Mollusks from Martha's Vineyard, Massachusetts," pp. 472-485 of the *Journal of Paleontology*, vol. 37, 1963.

G. E. Hutchinson shows how isotopic ratios in a fossil shell reveal the temperature of the water where its bearer lived in "Time, Temperature and Aphrodite," reprinted in his book, *The Itinerant Ivory Tower*, Yale University Press, 1953.

R. T. Paine discusses the significance of the tooth on the outer lip of some carnivorous snails in his paper, "Function of Labial Spines, Composition of Diet, and Size of Certain

Marine Gastropods," pp. 17-25 of *Veliger*, vol. 9, 1966.

Perhaps the best way to learn the relationship between an animal's teeth and its diet is to visit a museum.

J. W. Bews shows how to infer climate from leaf form in a series of papers, "Studies in the Ecological Evolution of the Angiosperms"; all of these are in vol. 26 of the *New Phytologist*, 1927.

A particularly good discussion and example of the methods of functional morphology as applied to fossils is M. J. S. Rudwick, "The Feeding Mechanism of the Permian Brachiopod *Prorichthofenia*," pp. 450-471 of *Palaeontology*, vol. 3, 1961.

Goulden discusses the application of MacArthur's broken stick model in his memoir, "La Aguada de Santa Ana Vieja: an Interpretative Study of the Cladoceran Microfossils," pp. 373-404 of *Archiv für Hydrobiologie*, vol. 62, 1966.

Stehli analyzes recent diversity gradients in "Taxonomic Diversity Gradients and Pole Location: the Recent Model," pp. 163-228 in E. Drake, *Evolution and Environment*, Yale University Press, 1968. He applies the same techniques to locate the position of the North Pole in Permian time in "A Paleoclimatic Test of the Hypothesis of an Axial Dipolar Magnetic Field," pp. 195-208 in R. A. Phinney, ed., *The History of the Earth's Crust*, Princeton University Press, 1968. The second paper is, however, only a brief progress report and the results it reports are very tentative.

CHAPTER 12

The Fossil Record

WHAT CAN we learn about evolution from the fossil record? How detailed an account can we reconstruct? What sort of theories can we expect to verify from fossils?

Fossils provide a narrative history of evolution: our first problem is to learn how to read the narrative. A fossiliferous sediment results from particles falling one upon another: they may be cinders raining down from a volcano or bits of sand and mud settling to the bottom of a lake or ocean. The newer particles fall atop the older, and so they remain unless a later disturbance folds the "depositional surface" over on itself. Thus we normally assume that if one fossil stratum overlies another, the upper is more recent. The fossil story always starts at the bottom, whether it be a few imprints of shells and corals in a road cut or the broad record exposed by the waters in the Grand Canyon.

While sediments form in basins, mountains erode away. Every locality has witnessed some erosion; there is no complete record of evolution in any one place, not even the Grand Canyon. We must correlate fragmentary records by the fossils they have in common, and synthesize from these a continuous narrative. A few sediments have radioactive minerals revealing their age: when fitted into the narrative they provide a time scale for it.

We correlate strata by tracing the consequences of worldwide events. The Paleozoic, Mesozoic, and Cenozoic are demarcated by events of unusual importance. The Paleozoic began with the appearance of marine shelled animals, especially trilobites. These were the first fossilizable animals to appear in abundance; they mark the start of a *coherent* fossil record. Why these animals appeared so suddenly was long a puzzle: it now seems the Cambrian was the first time sufficient oxygen was available to make respiration profitable. The earth began without an atmosphere: an atmosphere formed from volcanic gases containing methane, carbon dioxide, water vapor, and nitrogen, but no free oxygen. Oxygen would only accumulate when photosynthetic organisms appeared, and then only slowly. Oxygen was presumably most abundant on plant surfaces, and animals perhaps evolved as parasites attached to plants and dependent on them for food and oxygen. When the oxygen content of the atmosphere rose to about 1% of the present, permitting profitable respiration, the parasites evolved into the free-living animals which mark the Cambrian. As usual when a great opportunity opens, animals exploited their possibilities rather quickly.

Once respiration evolved, plants became far more efficient, and oxygen accumulated more rapidly. In the early Paleozoic, there was not sufficient oxygen to shield the earth against ultraviolet radiation, and many animals were shielded by exoskeletons. As oxygen increased, many of these exoskeletons were reduced or eliminated: fishes, which appeared in armor plate in the Ordovician, became their familiar scaly selves by the

Permian. Meanwhile, the land surface, uninhabitable when the Paleozoic began, became accessible to plants after oxygen had increased to a tenth its present level. Land plants, therefore, appeared much later than sea animals, and only began to multiply in earnest in the Silurian and Devonian.

The end of the Paleozoic was marked by a "crisis" of world-wide extinctions. The animals which suffered worst were those which today are most nearly restricted to waters of normal salinity: echinoderms, corals, brachiopods, cephalopods; this suggests that the crisis was caused by a drop in the salt content of the oceans. The crisis persisted through the lower Triassic: the marine communities of the time were composed of a few, widely distributed species, and many kinds of animals were absent which were abundant both before and after.

What could have lowered the salt content of the oceans? To reduce it by a seventh is to remove 750,000 cubic miles of salt. An answer is suggested by the Mediterranean, which loses more water by evaporation than it gains from rainfall. To make up the deficit, Atlantic water flows in. This displaces some of the saltier Mediterranean water, which, being denser, sinks. If the Mediterranean evaporated faster, its water would sink to the ocean floor, withdrawing its salt content from circulation. In the Permian, there were several nearly landlocked seas in regions of arid climate. We know that one of them removed ten thousand cubic miles of salt in 300,000 years (see Appendix). If deep sea brines mixed with surface waters at a rate high enough to restore normal salinity after twenty million years, it would have taken five such seas ten million years to remove a seventh of the oceans' salt. Since these brines lay so heavily on the ocean floor, phosphate-laden waters no longer welled up from the depths and the oceanic plankton starved.

The Mesozoic also closed with an unusual wave of extinctions. On land, the dinosaurs died out quite mysteriously, making way for the mammals. In the oceans, the plankton suddenly disappeared, wiping out many animals of the open sea, such as ammonoids, belemnites, and the seagoing reptiles. Coastal waters were less affected: the composition of coral reefs changed little, gastropods flourished more than ever, and clams survived with only a few extinctions. No one knows what caused the plankton failure (just as no one knows what caused the dinosaur failure), which wiped out nearly all the stocks of oceanic plankton. Perhaps deep sea brines once again prevented nutrients from leaving the depths.

Coral reefs vanished suddenly at the end of the Devonian, and did not reappear for forty million years; ammonoids died out at the end of the Triassic, leaving only a few survivors to repopulate the Jurassic seas. Such waves of extinction are rare, however: normally, we must correlate by the origin and spread of new groups of organisms. We can correlate deposits containing rapidly evolving, widely spreading animals to within a million years. Marine snails are dispersed so effectively by their planktonic larvae that a guide to the shells of Hawaii works surprisingly well in the Red Sea. (By contrast, forests so far apart share very few trees in common.) So with all animals living in the open sea, either as larvae or as adults. For some reason, many of them also evolve rapidly: these provide "index fossils" correlating their deposits. The graptolites, colonial animals which floated on Paleozoic oceans, correlate Ordovician and Silurian deposits; ammonoids correlate the Mesozoic. As they become better known, shelled planktonic protozoans correlate them all.

Glaciations and other widespread disturbances permit finer correlations, but they *disturb* the evolutionary process we wish to describe and understand. Our narrative inevitably blurs. The distribution of micrometeorites suggests that, even in apparently continuous deposits, sedimentation proceeds only a tenth of the time: it starts and stops, further blurring the narrative. Occasionally, we find more evenly sedimented deposits:

England's ammonite-bearing Oxford clay records, with few interruptions, the change in four species of ammonite over a million years. But such deposits (like those containing imprints of soft-bodied animals) are rare signposts allowing us to check our bearings along an otherwise ill-marked way.

The sea record blurs, but the land record distorts. The Devonian records mostly river-deltas; the Carboniferous, swamp-forests. A new appearance in this record may signal either the origin of a group or its migration from unrecorded highlands. A complete history is impossible: one cannot hope to explain fully every change in terms of its antecedents. The sketchy record is a severe but salutary discipline, forcing on us an economy of explanation commensurate with our limited knowledge. Unable to hide behind infinite detail, we must learn to identify and emphasize the important. Unable to construct a full narrative, we must settle for illustrated principles.

Appendix

A nearly landlocked sea in Texas and New Mexico, the Castile Sea, deposited 3.3×10^{12} tons of anhydrite ($Ca\ SO_4$) and rather less salt ($Na\ Cl$) in 306,000 years. How many cubic miles of salt did it remove from circulation during this period? (A cubic mile is 10^{10} tons.)

In normal sea water, there is thirty times as much salt as anhydrite: when this water is evaporated to a fifth its normal volume, the anhydrite begins to precipitate, but salt is deposited only after the water is less than a tenth its original volume. Thus, for every ton of anhydrite deposited, thirty tons of salt were returned to the ocean as concentrated brines which sank to its bottom. The Castile Sea thus removed 10^{14} tons, or 10,000 cubic miles, of salt from circulation during this period.

Bibliographical Notes

An elementary account of the fossil record and the means by which it is assembled is given in A.O. Woodford's *Historical Geology*, W. H. Freeman, 1966. There are few good historians of life. The best is G. G. Simpson, who studies a single group in depth in *Horses* (Oxford University Press, 1951), and discusses the fossil record as a whole in *Tempo and Mode in Evolution* (Columbia University Press, 1944) and *Major Features of Evolution* (Columbia University Press, 1953).

Oxygen's role in evolution is discussed in Berkner and Marshall's "History of Major Atmospheric Components," and A. G. Fischer's " Fossils, Early Life, and Atmospheric History," pp. 1215-1226 and 1205-1215, respectively, in vol. 53 of the *Proceedings of the National Academy of Sciences*, 1965: these papers are part of a symposium on the evolution of the earth's atmosphere.

For discussions of the "crises" of the fossil record, see A. G. Fischer, "Brackish Oceans as the Cause of the Permo-Triassic Marine Faunal Crisis," pp. 566-575 of A. E. M. Nairn, ed., *Problems in Paleoclimatology*, Interscience, 1964; M. N. Bramlette, "Massive Extinctions in Biotas at the End of Mesozoic Time," pp. 1696-1699 in *Science*, vol. 148, 1965; N. D. Newell, "Catastrophism in the Fossil Record," pp. 97-101 in *Evolution*, vol. 10, 1956; and N. D. Newell, "Paleontological Gaps and Geochronology," pp. 592-610 in *Journal of Paleontology*, vol. 36, 1962.

The earth's magnetic field reverses suddenly every few hundred thousand years, and the plankton extinctions associated with these reversals may permit a refined correlation: see Billy Glass and B. Heezen, "Tektites and Geomagnetic Reversals," pp. 33-38 in *Scientific American*, vol. 217, no. 1, 1967.

The ammonites of the Oxford clays are thoroughly analyzed in R. Brinkmann, "Statistische-biostratigraphische Untersuchungen an mitteljurassischen Ammoniten ueber Artbegriff und Stammesentwicklung," Abhandl. Ges. Wiss. Göttingen, Math. Nat. Klasse (N. F.), vol. 13, pp. 1-249, 1929. Although no one knows whether the four forms of ammonites whose evolution Brinkmann followed represent one, two, or perhaps even four species, his monograph is considered a classic case study of evolution. Another such study is D. Nichols, "Changes in the Chalk Heart-urchin *Micraster* interpreted in relation to Living Forms," *Philosophical Transactions of the Royal Society of London*, Series B, vol. 242, pp. 347-437, 1959. Unfortunately, this paper is so thick with necessary terminology that it is extremely difficult to read (imagine the strange terminology intelligent Martian starfish would have to invent to describe humans).

CHAPTER 13

How Predictable Is Evolution?

MADAGASCAR was an island before rain forest evolved. Its plants descend from ancestors which arrived before it was an island, and from seeds which blew or floated across the sea. Yet Madagascar has developed a tropical rain forest amazingly like rain forest anywhere. Although the trees are not so tall as their mainland counterparts (only 100 feet high, rather than 130 to 200), the trunks are, as usual, columnar, often buttressed at the base. The forest is too shady for much ground vegetation, and the visitor walks easily through the multitude of growing saplings, many of which are slender poles with only a few magnolia-shaped, dark green leaves at the top. He sees a pattern of rather pale tree trunks, grey, pink, orange, and brown, contrasting with the dark green foliage. The red of new leaves is more prominent than flowers. A look upward reveals a pattern of small sunlit leaves against the sky. Below them, often shaded, are occasional crowns of huge leaves belonging to palm or traveller's tree, and blankets of nearly foot-long leaves belonging to shrubs fifteen or twenty feet high. Sunlit branches bear woody vines (lianes) often thicker than a man's arm. Several thousand feet higher the forest is shorter and mossier, tree ferns more abundant. Branches are loaded with orchids, begonias, Kalanchoe, and the like: more flowers are visible. Even though this forest descended from weeds capable of dispersing across oceans, and evolved nearly isolated from its mainland counterparts, it looks very much the same as others of its kind.

How predictable is evolution? Does the origin of life imply the evolution of intelligence? Would another planet "just like earth" have flowering trees with green leaves, insects on the flowers, monkeys on the branches, and tigers stalking the monkeys? Or would we find beings resembling trees, monkeys, etc. yet fundamentally different biochemically? Or would we find the forms as bizarre as the biochemistry? Comparison of mainland with island evolution provides our only clues to such questions. Each island provides a different stage for evolution, so comparison also allows us to judge the effects of different factors on evolution.

On the Galápagos, Darwin's finches have filled many roles, but their evolution is so incomplete that their close family resemblance betrays their common ancestry. Some are rather strange: the woodpecker finch picks out his insects with a thorn. Mainland woodpeckers are undoubtedly more efficient, but this finch is caught in an evolutionary "blind alley." Why does he survive? Are the islands too small for another woodpecker to evolve? Is there no opportunity to test different woodpeckers against each other?

Aside from the bats, Madagascar's mammals derive from four colonists: a curiously archaic insectivore, a lemur, a rat, and a primitive civet not unlike a weasel. The insectivore's descendants (Fig. 13-1) resemble the moles, shrews, and hedgehogs of Europe, and, in general, the small mammals are much like those anywhere. The evolution of the middle-sized mammals is

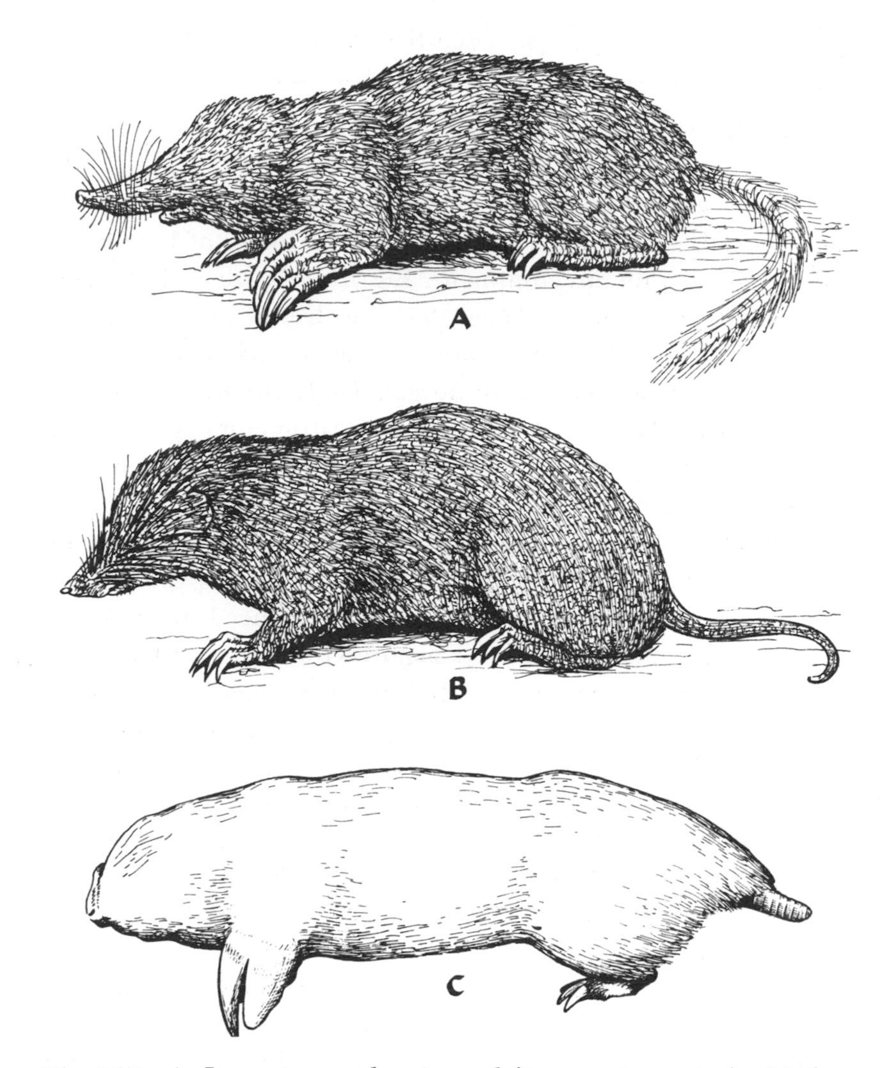

Fig. 13-1. A Japanese mole, A, and its counterparts in Madagascar, B, and Australia, C. (After Figs. 1660 and 1639A of H. Heim de Balsac and F. Bourlière, "Ordre des Insectivores: Systematique," in P. P. Grassé, *Traité de Zoologie*, tome XVII, fascicule II, pp. 1689 and 1662, and Fig. 135A of F. Bourlière, "Ordre des Marsupiaux; Systematique," in Grassé, *op. cit.*, tome XVII, fascicule I, p. 153. Used with permission of Masson et Cie., Paris.)

less complete: lemurs are less intelligent than the monkeys of continental jungles. Some, moreover, are rather unusual: the aye-aye, a rodent-toothed lemur which probes rotten logs for insects with a curiously prolonged fourth finger of bare bone, has no parallel on any continent. Big animals are entirely lacking: there is no carnivore larger than a wolverine, and the largest herbivore is a lemur. There are no grazing mammals. Indeed, before man came, there were no grasslands either, even though much of the island now resembles the cattle country of southeastern Wyoming. Were the native herbivores too inefficient for grassland to develop?

Australia was settled by a marsupial and, later on, by island-hopping mice. Despite the imbalance among the original colonists, its mammal fauna is more balanced than Madagascar's: it includes grazers (kangaroos) and a wolf-like carnivore (the Tasmanian wolf). Before man came, there were rhinoceros-like herbivores, ten-foot kangaroos, and a lion-sized carnivore. The rain forest of northern Australia shelters a marsupial analogue of the aye-aye, complete with prolonged fourth finger and rodent incisors. Like the aye-aye, it taps on wood to locate insects which it then scrapes out with its finger or its teeth. Are such animals characteristic of the early stages of mammal evolution, or is the parallel an accident? A "lemur" (the cuscus) has also evolved, but it is far less intelligent than its Malagasy counterparts. Marsupials are all rather stupid. Is the cuscus stupid because its ancestors were, or is it as smart as it needs to be to escape its predators? Has the evolution of its intelligence been cramped by its ancestry, or by lack of space?

South America became an island just as mammals were diversifying, and remained so until a few million years ago. Its mammals derived from aboriginal marsupial carnivores and placental herbivores, and from rats and lemurs invading during the Oligocene. It evolved a balanced fauna (Figs. 13-2 and 13-3), with "wolves" and "saber-tooth tigers" among the marsupials and "horses," nine-foot armadillos, and primitive "elephants" among the placentals. The lemurs gave rise to monkeys, some of whose modern descendants can express a very human range of emotions. Could South America evolve such intelligent monkeys while still an island? Or did these monkeys evolve only after relatively intelligent placental carnivores replaced the marsupials? The fossil record is still unclear on this point, but a few fossil brain cases might tell us a lot.

A few million years ago, in the late Pliocene, the isthmus of Panama formed, and the mammals of North and South America began to mix. During the Cenozoic, North America was frequently connected with Asia. Although monkeys and other jungle animals never crossed, Siberia and Alaska were usually warm enough to permit the exchange of temperate-zone

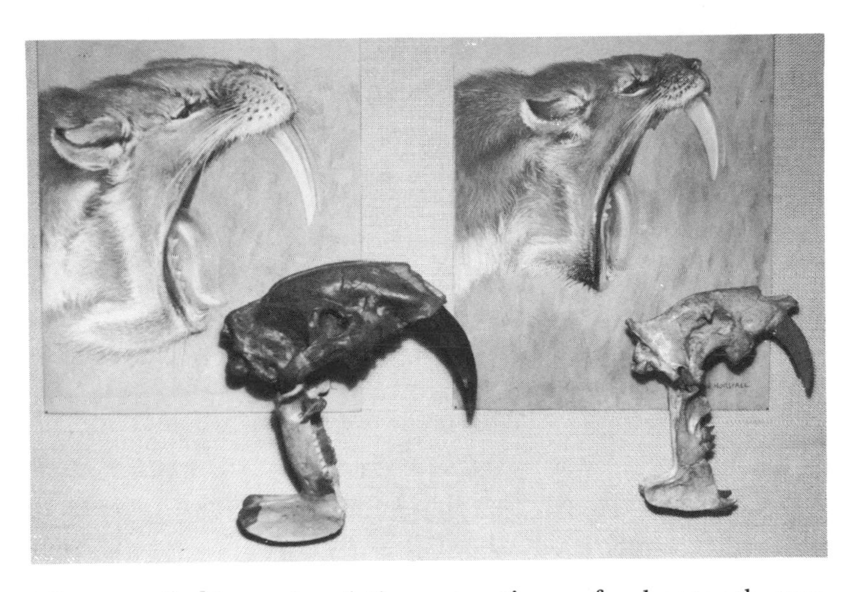

Fig. 13-2. Skulls and artist's restorations of saber-tooth marsupial *Thylacosmilus atrox* (Pliocene of Argentina) and saber-tooth cat *Eusmilus sicarius* (Oligocene of South Dakota). (Photograph by D. Baird.)

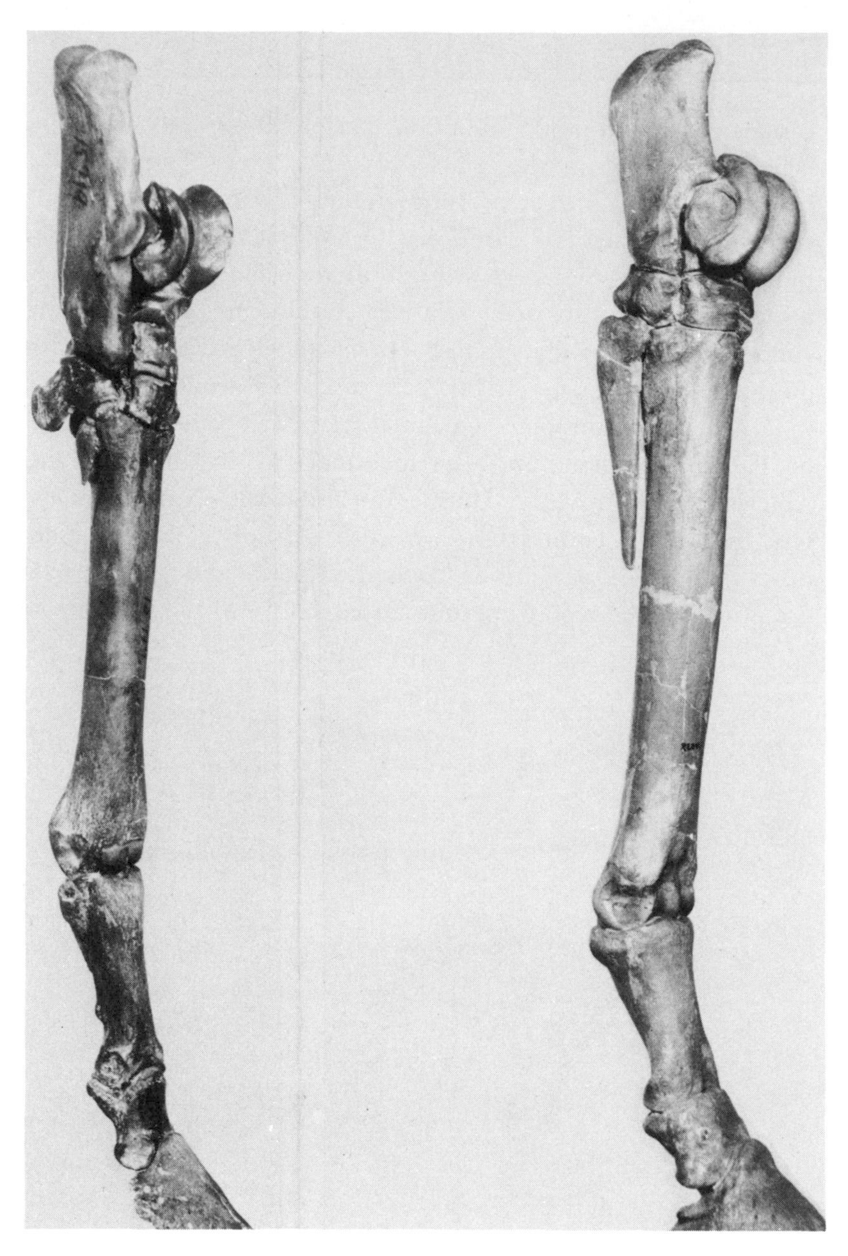

Fig. 13-3. A. Right hind foot (195 mm. long) of pseudo-horse *Thoatherium minusculum* (Miocene of Patagonia).
B. Right hind foot (545 mm. long) of true horse *Equus* (Pleistocene of Nebraska). (Photographs by D. Baird.)

animals. In effect, North America was part of a "world continent." Large as South America was, its mammals were less well adapted than their North American counterparts: when Panama formed, all its marsupial carnivores and hoofed herbivores were replaced. Armadillos and sloths survived because they had no counterparts to the north: even the "world continent" was not large enough to guarantee filling all possible mammal roles.

Islands tell us that the larger the landmass, the more predictable evolution is: where there is space enough, small mammals will eventually evolve wolves and horses, and perhaps also an intelligence to enjoy them. Natural selection depends for its effectiveness on a series of chances, the organ developed for one purpose being suitable for another, etc. Our ear mechanism includes bones which hinged the jaws of our reptilian ancestors; our thyroid is made from an embryonic gill-slit. The larger the landmass, the greater the opportunity for the successions of events required to fill the different ecological roles. Conversely, a small landmass cramps evolution: old islands are reliquaries of primitive forms.

What order there is in evolution results, however, from selection among the products of accident. We can expect only the broadest features of evolution to be predictable. Taking a longer view, we have no idea how different today's world would be had the accidents creating life turned out differently. Was a fundamentally different outcome possible? The question only reveals our ignorance.

Bibliographical Notes

D. Lack discusses the evolution of the Galápagos finches in his book, *Darwin's Finches*, Cambridge University Press, 1947.

G. G. Simpson discusses the ancestries of Australian and South American mammals (and how he deduced them!) in his book, *The Geography of Evolution*, Chilton, 1965. P. Darlington summarizes information on the ancestry of different faunas in *Zoogeography*, Wiley, 1967. An explicit comparison of mammal evolution in Africa, Australia, and South America is given in an excellent symposium organized by A. Keast, "Evolution of Mammals on Southern Continents," published in volumes 43 (1968) and 44 (1969) of the *Quarterly Reviews of Biology*. See especially A. Keast, "Introduction: the Southern Continents as Backgrounds for Mammalian Evolution," pp. 225-233 in vol. 43; A. Keast, "Australian Mammals: Zoogeography and Evolution," pp. 373-408 in vol. 43; B. Patterson and R. Pascual, "The Fossil Mammal Fauna of South America," pp. 409-451 of vol. 43; and A. Keast, "Comparison of the Contemporary Mammalian Faunas of the Southern Continents," pp. 121-167 of vol. 44.

A first attempt to compare communities of similar setting in different regions is Hesse, Allee, and Schmidt, *Ecological Animal Geography*, Wiley (second edition, 1951). See also M. L. Cody's precise comparison of grassland birds, in his papers, "The Consistency of Intra- and Inter-continental Grassland Bird Species Counts," pp. 371-376 of the *American Naturalist*, vol. 100, 1966, and "On the Methods of Resource Division in Grassland Bird Communities," pp. 107-147 in the *American Naturalist*, vol. 102, 1968. D. Lack presents another such comparison in his paper, "Tit Niches in Two Worlds; or, Homage to Evelyn Hutchinson," pp. 43-50 in the *American Naturalist*, vol. 103 (1969).

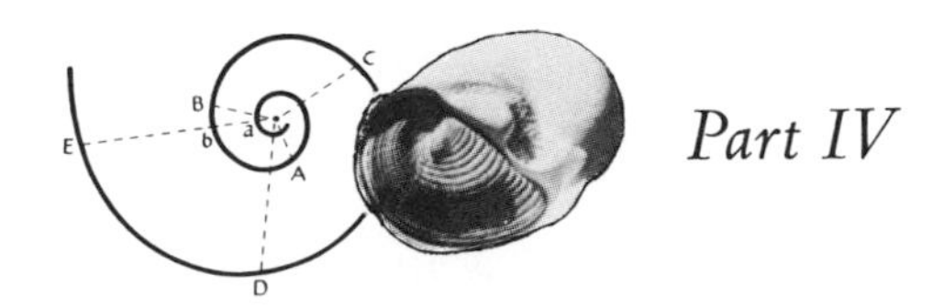

Part IV

THE MECHANISMS OF EVOLUTION

Finally, having set our enquiry in perspective, we can ask how organisms are adapted to their ways of life. To what extent does evolution favor the good of the species over the individual's advantage? To what extent does evolution favor the long-term good of a species over its momentary advantage? What are the mechanisms involved?

CHAPTER 14

Population Genetics

IN A sense, the characteristics of an organism are determined by its genes: in the same sense, the characteristics of a population are determined by its spectrum of gene combinations. The population evolves when its genetic composition changes. What can change its genetic composition?

Consider a single locus, with alleles A and a, in a population of diploids.* Suppose the population does not receive migrants from outside. Suppose, moreover, that successive generations are distinct (as in annual plants): all members of one generation

---

* A locus is a part of a chromosome concerned with a specific function: one speaks of the locus for insulin, or a locus for eye color. A haploid individual has one gene at each locus; a diploid, two. The genes at a locus belong to one or more alleles: genes belong to the same allele if they program the same instructions. At the eye color locus, one allele programs blue eyes, another brown, etc. A diploid individual is homozygous for eye color if both its genes at this locus are the same allele: otherwise, it is heterozygous.

die before any of their offspring mature. The *frequency* of *A* is the proportion of genes at this locus which are *A*: this is the frequency of *AA* homozygotes, plus half the frequency of the heterozygotes. If *A*'s frequency is q, then *a*'s is $1 - q$. The genes carried by one generation are a sample from the gametes of their parents: the frequency of *A* changes only if the sample is biased by mutation or natural selection, or if the sample is too small to reflect adequately *A*'s frequency in the previous generation.

Changes in genetic composition need not involve changes in allele frequencies. Consider a second locus, with alleles *B* and *b*, on the same chromosome as the first. If the population has chromosomes bearing *A* and *b*, *a* and *B*, and *a* and *b*, but none with *A* and *B*, it cannot form the gene combinations (genotypes) *AABb*, *AaBB*, or *AABB*. An *Ab* chromosome must exchange segments with an *aB* (in other words, "cross-over" must occur between an *Ab* and an *aB* chromosome) to form *AB* and *ab* before these genotypes can form, but cross-over does not change allele frequencies. Sexuality permits such recombination: this so facilitates evolution that most populations are sexual.

In sum, a population's genetic composition remains unchanged only if allele frequencies do not change and if cross-over forms each genotype as often as it destroys it. The only causes of evolution are therefore natural selection, mutation, sampling error, and recombination. Natural selection is the only one of these which adapts populations to their environments. The perfection of living things can hardly be ascribed to the blind reshuffling of genes or to the accidents of sampling error. We proofread books, because misprints rarely improve them: likewise, special enzymes proofread copies of genes, to correct mutations. The role of mutation is to create, and of sampling error and recombination to arrange, the variation on which natural selection acts.

How strong are the factors of evolution? Genes mutate

rather rarely. Mothers of hemophilia-free descent bear hemophiliac children but once in fifty thousand births. This rate is high: among fruit flies, perhaps one gene in a million mutates *detectably* each generation; among bacteria, perhaps one in a hundred million. Mutation pressure, therefore, requires thousands or millions of generations to appreciably alter allele frequencies. Such weak pressures rarely have a visible effect.

Consider a locus with alleles B and b. If each mutates to the other, mutation pressure shifts allele frequencies toward an equilibrium where as many genes mutate from B to b as vice versa. At equilibrium, the number of B-genes, times the proportion of these mutating to b in one generation, is equal to the corresponding product for b: the ratio of B's frequency to b's is the ratio of b's mutability to B's. Mutational equilibria have been demonstrated in laboratory cultures of bacteria, but none have been found in any natural population. For the mathematics of such equilibria, let B's frequency in generation t be q_t; let the proportion of B-genes miscopied as b (the proportion of B-genes mutating to b) each generation be u, and let v be the corresponding "mutation rate" from b to B. If mutation pressure is the only influence on B's frequency, then the frequency q_{t+1} of B in generation $t + 1$ will be that in generation t, less an amount uq_t due to mutation from B to b, and augmented by an amount $v(1 - q_t)$ by mutation from b to B. Thus

$$q_{t+1} = q_t - uq_t + v(1 - q_t) = q_t(1 - u - v) + v.$$

The equilibrium frequency q_e is that for which mutation from B to b just balances that from b to B: thus $v(1 - q_e) = uq_e$ or $q_e = v/(u + v)$. Substituting $v = q_e(u + v)$ into the equation for q_{t+1}, we obtain

$$q_{t+1} = q_t(1 - u - v) + (u + v)q_e.$$

Subtracting q_e from both sides of the equation, this becomes

$$q_{t+1} - q_e = (1 - u - v)(q_t - q_e).$$

The difference between B's frequency and its equilibrium declines by a factor $1 - u - v$ per generation: if u and v be small, this approach to equilibrium will be very slow.

A gene is a succession of "codons" which programs a corresponding sequence of amino acids in a protein. Many codons can be miscopied without visible harm: a surprising number of a protein's amino acids can be substituted without altering its shape or function. The fate of such an "invisible mutant" is governed by sampling error. If it is selectively neutral it will, on the average, leave one copy of itself in every future generation: that is to say, if a large number of "identical" populations are started with one neutral mutant apiece, there will always be as many mutants as populations. In each population, however, the mutant's numbers will "drift" randomly, according to the accidents of sampling, until either the mutant or its alternative allele disappears from the population (Fig. 14-1). If each population has $2N$ genes per locus, the mutant will entirely replace its allele in one out of every $2N$ populations, and will disappear from all the rest. The number of mutants per gene is the same as the number of alleles substituted per population. One of the amino acids of hemoglobin is substituted every ten million years or so in every vertebrate species, no matter how large its population or how fast its visible characters may be changing. Are these substitutions accidents of sampling? This implies that hemoglobins of different animals are equally mutable, and that each year a codon mutates in one of every ten million hemoglobin genes.

How quickly can sampling error alter an allele's frequency? Consider a locus with alleles B and b in a population of N diploid individuals. If B begins with frequency 1/2, how long before it either replaces b or disappears from the population? Suppose the population produces abundant gametes, from which $2N$ are chosen at random to form the next generation. Let F_t be the probability that, in generation t, two genes chosen

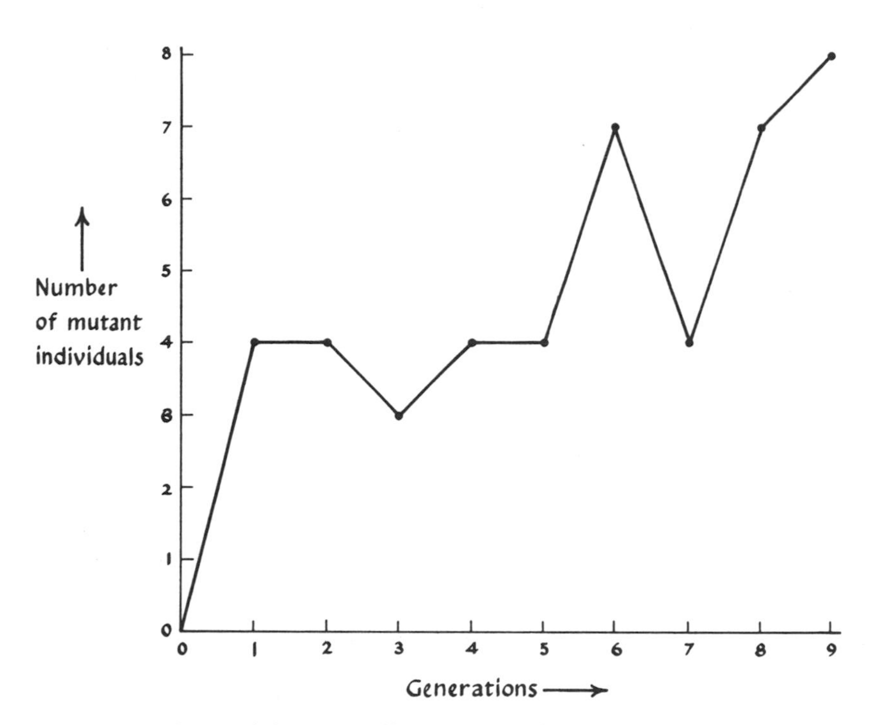

Fig. 14-1. Spread by sampling error of a recessive mutant in a population of constant size 8, from data of S. Wright. Graph records the number of individuals of mutant *phenotype* in successive generations.

at random from the same locus are the same allele. In generation 0, when each allele's frequency is 1/2, F is the probability $(1/2)^2$ that both genes are B, plus the probability, again $(1/2)^2$, that both genes are b, or 1/2. When one of the alleles has attained the frequency $1 - z$, where z is a small number, then $F = (1 - z)^2 + z^2$, which is very nearly 1. How quickly does F approach 1 (Fig. 14-2)? The probability F_t that, in generation t, two genes are the same allele is the probability $1/2N$ that they were copied from the same parent gene, plus the probability

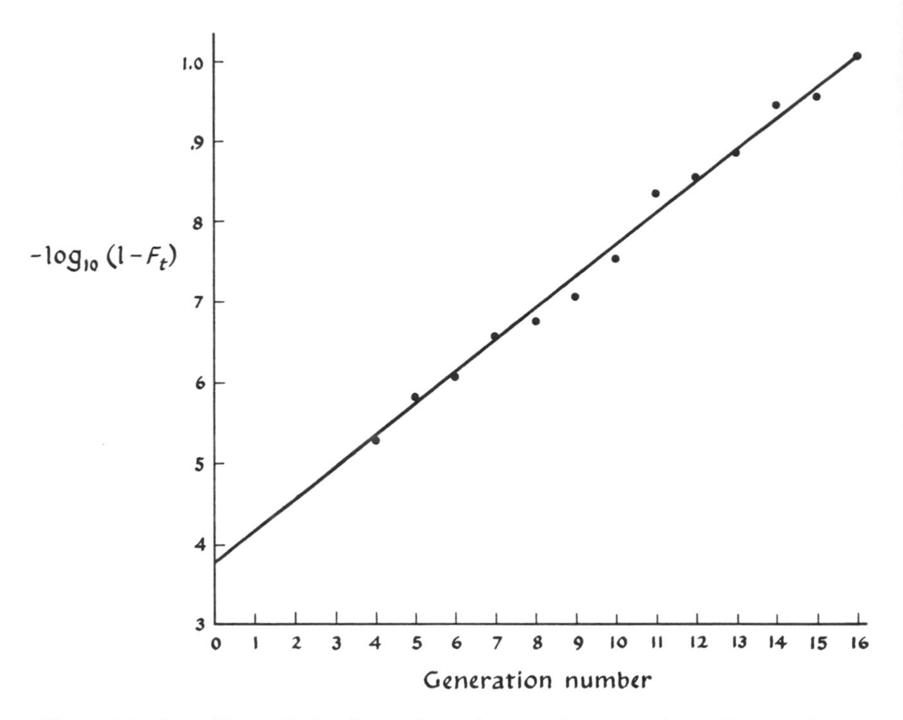

Fig. 14-2. log $(1 - F_t)$ plotted against t for a series of populations of size 8: F_t is the probability that two genes drawn at random from the same population t generations after the experiment began will be the same allele. Data from Kerr (see Problem 1). The line is the one best fitting the data for generations 4-16.

$(1 - 1/2N)F_{t-1}$ that they were copied from different genes which were still the same allele. In symbols,

$$F_t = 1/2N + (1 - 1/2N)F_{t-1},$$

$$1 - F_t = (1 - 1/2N)(1 - F_{t-1}) =$$

$$(1 - 1/2N)^t(1 - F_0) \cong (1/2)e^{-t/2N}.$$

Since most species number millions of individuals, this theory suggests that sampling error is as weak an influence as mutation pressure.

The theory, however, is more suggestive than convincing. Sampling error hinders a local population's adaptation to special features of its habitat, even when it fails to influence the species as a whole. Moreover, its influence is often far stronger than a population's numbers suggest. A population a hundredth of whose members raise a hundred offspring each while the others remain childless is as susceptible to sampling error as a human population a hundredth its size. Many animals do live or die as whole broods. A butterfly may lay many eggs in one place: if a predator chances on them, he will eat them all; and if bad weather kills any, it will probably kill all. One textbook on evolution describes a swarm of six-legged frogs in a Mississippi pond, which must have come from a single lucky brood: sampling error could not spread a mutant so quickly in a human village.

Moreover, sampling error each generation is proportional to that generation's numbers, so sampling error is primarily governed by the population's minimum size. For example, the cumulative sampling error in a population whose successive generations number 1,000, 10,000 and 100,000 is proportional to

$$\frac{1}{1000} + \frac{1}{10000} + \frac{1}{100000} \cong \frac{3}{2700}.$$

This population is as susceptible to sampling error as one of constant size 2700. Many populations do cycle: Canada's lynxes are twenty times as common in good years as in bad, and the hares they eat fluctuate even more. The occasional population crash has an effect seemingly out of all proportion to its frequency: a population numbering a million for forty-nine generations and a thousand the fiftieth suffers as much sampling error as one of constant size 50,000. Who can guess the frequency or extent of such catastrophes?

Natural selection occurs when one allele is more reproductive than another. The frequency of the more reproductive allele obviously increases, but how fast? Consider a locus with alleles *B* and *b* in a population of haploids (which have one gene at each locus). The frequency of *B*, which we call q, is simply the proportion of individuals in the population which carry *B*. If each *B*-gene passes on W_B copies of itself to the next generation, while *b*-genes pass on the smaller number W_b, the ratio of *B*'s numbers to *b*'s (which is the ratio of *B*'s frequency to *b*'s) multiplies by W_B/W_b each generation: it changes by geometric progression. The proportion by which W_B exceeds W_b is called the selective advantage s of *B* over *b*: in symbols, $W_B/W_b = 1 + s$. The ratio $q_t/(1 - q_t)$ of *B*'s frequency to *b*'s in generation t is $1 + s$ times that in generation $t - 1$, or $(1 + s)^t$ times that in generation 0: in symbols,

$$\frac{q_t}{1 - q_t} = \frac{(1 + s)q_{t-1}}{1 - q_{t-1}} = (1 + s)^t \frac{q_0}{1 - q_0} \cong e^{st} \frac{q_0}{1 - q_0}.$$

B's frequency takes as long to change from 1/900 to 1/100, or from 99/100 to 899/900, as from 1/4 to 3/4: the change in gene ratio is the same in all three cases. Selection is thus most effective on alleles with frequencies close to 1/2: it is the less effective the closer an allele's frequency to either 0 or 1.

In striking contrast to mutation pressure or sampling error, selection can rapidly transform a natural population: the speed with which mosquitoes evolve resistance to DDT is just one example. Yet, rapidly as natural selection sometimes works, it can lead to very precise adaptation. Chapter 3 showed that, among humans, selection favors a slight excess of male births to compensate the higher mortality among male children. This selection is very weak: if the optimum sex ratio at birth is 105 boys to 100 girls, an allele which, when homozygous, causes a sex ratio of 108 : 100 has a selective disadvantage of only one in twenty thousand (this is shown in the appendix). Yet the ratios of boys to girls born in different countries nearly all lie between 102:100 and 108:100. Is adaptation even more precise than this? Have these ratios evolved to suit the mortality patterns of different races? We may never know: sex ratio evolved in primitive man, with death rates very different from today's.

What of selection in diploids? Consider a locus with alleles *B* and *b*. If *B*-genes pass on W_B copies for each parent gene, and *b*-genes W_b, the ratio of *B* to *b* still multiplies by W_B/W_b per generation. However, a gene's reproductivity depends on whether it occurs in a heterozygote or a homozygote: this means, as we shall see, that the gene ratio rarely changes according to a strict geometric progression. If the heterozygote is more reproductive than either homozygote, the frequency of a rare allele would increase, since it would occur mostly in heterozygotes. Selection thus maintains both alleles in the population, and the gene ratio tends toward an equilibrium where the alleles are equally reproductive, where the selection against one allele's homozygotes exactly balances that against the other's. Such alleles comprise a "balanced polymorphism." The "sickle-cell" polymorphism of West African Negroes is perhaps the most famous of the many balanced polymorphisms discovered so far: the sickle-cell allele immunizes its heterozygotes against an often fatal malaria, but its homozygotes usually die of anemia. Balanced polymorphisms allow a population to respond to environmental change without losing the capacity to adapt: haploids are at something of a disadvantage because they cannot form such polymorphisms.

To describe precisely selection in diploids, we must calculate the average reproductivities of the alleles *B* and *b* in terms of those of the genotypes *BB*, *Bb*, and *bb*. If a *B*-gene carried in a homozygote has W_{BB} "children," while one in a heterozygote has W_{Bb}, the average number W_B of children per parent *B*-gene is W_{BB} times the proportion of *B*-genes in homozygotes, plus W_{Bb} times the proportion in heterozygotes. If the members of our population mate at random, a proportion q of the B-genes will be carried in homozygotes*(a proportion q of the individuals carrying B on one chromosome will carry it on its homologue as well), so $W_B = qW_{BB} + (1-q)W_{Bb}$. Similarly, $W_b = qW_{Bb} + (1-q)W_{bb}$. The ratio $q_t/(1-q_t)$ of B's frequency to b's in generation t is related to that in generation $t-1$ by

$$\frac{q_t}{1-q_t} = \frac{W_B}{W_b}\frac{q_{t-1}}{1-q_{t-1}} = \left[\frac{W_{BB}q_{t-1}+W_{Bb}(1-q_{t-1})}{W_{Bb}q_{t-1}+W_{bb}(1-q_{t-1})}\right]\frac{q_{t-1}}{1-q_{t-1}}.$$

Suppose the heterozygote is more reproductive than either homozygote, and set $W_{BB} = W_{Bb}(1-K)$, $W_{bb} = W_{Bb}(1-k)$: K and k are the selective disadvantages of BB and bb relative to Bb. Then

$$\frac{q_t}{1-q_t} = \left[\frac{q_{t-1}(1-K)+1-q_{t-1}}{q_{t-1}+(1-q_{t-1})(1-k)}\right]\frac{q_{t-1}}{1-q_{t-1}}$$

$$= \left[\frac{1-Kq_{t-1}}{1-k(1-q_{t-1})}\right]\frac{q_{t-1}}{1-q_{t-1}}.$$

q increases if $Kq < k(1-q)$ and decreases otherwise: B's frequency thus approaches an equilibrium q_e such that $Kq_e = k(1-q_e)$, or $q_e/(1-q_e) = k/K$.

If the members of a population mate at random, recombination associates alleles from different loci at random. If allele frequencies are not changing, the population approaches an

---

* Consider the human blood-group locus with the alleles M and N. In one population, the frequencies of MM, MN, and NN were .2838, .4957, and .2245, respectively. Is the probability of an M-gene being in a homozygote equal to M's frequency?
Solution: The probability of an M-gene being in a homozygote is .2838/(.2838 + .4957/2),
or .534. M's frequency is .2838 + .4957/2, or .532. These numbers do not differ significantly.

equilibrium where the frequency of each genotype is equal to the product of the frequencies of its component alleles. Consider, for example, the effect of recombination between the loci A, a and B, b. Let the frequencies in generation t of chromosomes bearing A and B, A and b, a and B, and a and b be $P(t)$, $Q(t)$, $R(t)$, and $S(t)$, respectively. $P(t) + Q(t)$ is the frequency q of B; $P(t) + R(t)$ is the frequency of A, which we call p. If a proportion k of these chromosomes "cross over" between these loci each generation, then the frequency $P(t+1)$ of AB chromosomes in generation $t+1$ is the frequency $(1-k)P(t)$ of those inherited intact from generation t, plus the frequency kpq of AB chromosomes formed by recombination. In symbols,

$$\begin{aligned}
P(t+1) &= (1-k)P(t) + kpq; \\
P(t+2) &= (1-k)P(t+1) + kpq \\
&= (1-k)^2 P(t) + [1-(1-k)^2]\,pq; \\
P(t+n) &= (1-k)^n P(t) + [1-(1-k)^n]\,pq.
\end{aligned}$$

The frequency of AB chromosomes approaches ever more closely to pq. Recombination usually assures that an allele is tested on its own merits: it cannot spread simply by association with other genes.

Appendix – Selection for Sex Ratio

Consider a locus affecting sex ratio, with alleles A and a (see appendix to Chapter 3). Define the frequency of A to be $\frac{1}{2}(q_m + q_f)$, where q_m is its frequency among the population's males and q_f its frequency among the females. Suppose all genotypes are equally reproductive, so selection favors a 1:1 ratio; suppose also that, on the average, A-bearers produce $1 + K_A$ sons and $1 - K_A$ daughters apiece, while the averages for the population as a whole are $1 + \overline{K}$ and $1 - \overline{K}$. How much will A's frequency change from one generation to the next?

If A's frequency is q in one generation, in the next it will be

$$\frac{1}{2}\left\{\frac{(1+K_A)q}{1+\bar{K}}+\frac{(1-K_A)q}{1-\bar{K}}\right\}=\left\{\frac{1-K_A\bar{K}}{1-\bar{K}^2}\right\}q.$$

If $\bar{K}^2$ is small compared to 1, A's frequency changes by the amount $q\bar{K}(\bar{K}-\bar{K}_A)$ in one generation.

To be specific, let us calculate a's selective advantage if a family's sex ratio is determined by the father's genotype, and if AA fathers have one son and one daughter apiece, while the ratios for Aa and aa fathers are 1.0075:.9925 and 1.015:.985, respectively. If q_m is A's frequency among parent males, the ratio of sons to daughters for A-bearing fathers is $1.000q_m + 1.0075(1-q_m){:}1.000q_m + .9925(1-q_m)$, or $1+.0075(1-q_m){:}1-.0075(1-q_m)$. The sex ratio among all the population's offspring is

$$1.000q_m{}^2+1.0075(1-q_m)2q_m+1.015(1-q_m)^2:$$
$$1.000q_m{}^2+.9925(1-q_m)2q_m+.985(1-q_m)^2.$$

This simplifies to $1+.015(1-q_m){:}1-.015(1-q_m)$. The sex ratio among the offspring of A parents is the proportion of these offspring (approximately 1/2) born of A fathers, times the ratio of sons to daughters for A fathers, plus the proportion of these offspring born of A mothers, times the sex ratio for the population as a whole (the mother's genotype is assumed not to influence family composition). Thus,

$$1+K_A=1+.015(1-q_m)-½(.0075)(1-q_m);$$
$$1+\bar{K}=1+.015(1-q_m).$$

The change in A's frequency over one generation is $q\bar{K}(\bar{K}-\bar{K}_A)$, or

$$½(.015)(.0075)(1-q_m)^2q.$$

This is approximately $.000056\ (1-q_m)^2$. Were a sex ratio of

105:100 optimum, it would be similarly favored over a ratio of 108:100 (or 1.065:.985).

Problems

1. The frequency of the sex-linked allele "forked" changes at random in small populations. To assay the effects of sampling error, Kerr formed 96 populations. Each generation of each population consisted of four males and four females drawn at random from the offspring of the generation previous. "Forked" had frequency 1/2 in the parental generations of all these populations. In successive generations of offspring, Kerr recorded the following number of populations still containing both alleles:

generation no.	4	5	6	7	8	9	10	11	12	13	14	15	16
no. of unfixed populations	79	70	66	59	56	52	47	39	37	34	30	29	26

(Data from Kerr and Wright: "forked" was apparently selected against after generation 11.)

a) Calculate F_t, the probability that two genes drawn at random from the same population are the same allele, for generations 4-16.

Since females have two genes apiece at this locus, and males only one, each population has 12 genes. Assume that in the populations still containing both alleles, all frequencies are equally probable: i.e., one-eleventh of these populations have one forked gene, one-eleventh have two, etc.: in a population of N haploids, this assumption is quite accurate after $N/2$ generations.

b) If "forked" is subject only to sampling error, $1-F_t = (1 - F_0)(1 - \alpha)^t$. What value of α best fits the data? $1/\alpha$ is called the "effective population number," since a population of this many haploids would behave as these do.

Solution:

a) If n unfixed lines remain at generation t, then $F_t = (96 - n)/96 + nP/96$, where P is the probability of two genes being the same allele in a population still containing both alleles. P is the sum

$$(1/144 + 121/144)/11 + (4/144 + 100/144)/11 + (9/144 + 81/144)/11 + \ldots,$$

or 23/36. $1 - F_t = (n/96)(1 - P)$, or $(n/96)(13/36)$. The values of $1 - F_t$ in successive generations t are as follows:

t	0	4	5	6	7	8	9
$1 - F_t$	.500	.297	.263	.248	.222	.211	.196

t	10	11	12	13	14	15	16
$1 - F_t$	.177	.147	.139	.128	.113	.109	.098

b) If $1 - F_t = (1 - F_0)(1 - \alpha)^t$, then $\log(1 - F_t) = \log(1 - F_0) + t\log(1-\alpha)$. $\log(1 - F_t)$ is graphed against t in fig. 14-2; the data are best fitted by the line

$$\log_{10}(1 - F_t) = -.3645 - .0405t.$$

If $\log(1 - \alpha) = -.0405$, $1 - \alpha = .9110$, $\alpha = .0890$, and the effective population number $1/\alpha$ is 11.2. Notice that, according to this equation, $1 - F_0 = .43$: such a deviation is reasonably ascribed to chance. (If Kerr had repeated his ensemble of experiments a hundred times and an equation fitted to $\log(1 - F_t)$ each time, seven of the values of $1 - F_0$ calculated from these equations would differ as much from .5.)

2. The normal form of the British peppermoth, *Biston betularia*, is colored to blend with lichen-covered tree trunks. Industrial smoke has covered many British woodlands with soot, against which these light-colored moths are all too visible: here, a melanic form is favored. This moth has one generation a year. Assume the color locus has two alleles, the normal a and the melanic A, with the following properties:

1) a mutates to A at the rate $u = 10^{-5}$.

2) In normal environments, homozygous A-genes, genes in heterozygotes, and homozygous a-genes produce W, $W\sqrt{2}$, and $2W$ offspring per parent, respectively, while in sooty places, the advantages of A and a are reversed.

a) What is A's equilibrium frequency in normal environments?

b) How many generations of sooty conditions are required to increase A's frequency to .8?

Solutions:

a) If A's frequency is q in generation t, then in the absence of mutation it will be

$$\frac{Wq^2 + W\sqrt{2}q(1-q)}{Wq^2 + 2W\sqrt{2}q(1-q) + 2W(1-q)^2}$$

$$= \frac{q[q+\sqrt{2}(1-q)]}{[q+\sqrt{2}(1-q)]^2} = \frac{q}{q+\sqrt{2}(1-q)}$$

in generation $t + 1$. A's frequency is in equilibrium if the increase due to mutation from a precisely balances the decrease from selection; that is, if

$$q = q/[q+\sqrt{2}(1-q)] + 10^{-5}(1-q),$$

q will be small, so $10^{-5}(1-q)$ will be nearly 10^{-5} and

$q/[q + \sqrt{2}(1-q)]$ will be approximately $q/\sqrt{2}$. Thus the condition of balance is very nearly

$$q = q/\sqrt{2} + 10^{-5},$$

$$q = 10^{-5}/(1 - \sqrt{1/2}\,), \text{ or nearly } (1/3) \times 10^{-4}.$$

b) In a sooty environment, the gene ratio $q/(1-q)$ changes by a factor

$$\frac{2Wq + \sqrt{2}\,W(1-q)}{\sqrt{2}\,Wq + W(1-q)} = \sqrt{2}$$

each generation. When $q = (1/3) \times 10^{-4}$, $q/(1-q) = (1/3) \times 10^{-4}$; when $q = 4/5$, $q/(1-q) = 4$. The time required to shift the gene ratio from $(1/3) \times 10^{-4}$ to 4 is given by the equation

$$(1/3) \times 10^{-4} (\sqrt{2}\,)^t = 4.$$

Taking logarithms of both sides and rearranging, we find that 34 generations (34 years) are required for such a change.

In actual fact, the melanic gene evolved overdominance during its spread, so that heterozygotes are now favored over both homozygotes, and the "normal" form is never quite eliminated.

Bibliographical Notes

Population genetics was shaped by three fundamental publications: R. A. Fisher, *The Genetical Theory of Natural Selection*, Oxford, 1930 (revised ed. Dover, 1958); J. B. S. Haldane, *The Causes of Evolution*, Longmans Green, 1932 (reprinted by Cornell University Press, 1966); and

S. Wright, "Evolution in Mendelian Populations," pp. 97-159 of *Genetics*, vol. 16, 1931 (reprinted in S. Wright, *Systems of Mating and Other Papers*, Iowa State College Press at Ames, 1958). Fisher's work is an enduring classic: one sometimes feels it was constructed as a marvelous and impenetrable diamond. Its difficulty was largely due to Fisher's originality as a mathematician; mathematicians are still learning how to use his tools. The later chapters of the book are a curiously anachronistic essay on the relative infertility of the upper classes and the misfortune of Britain's declining birth rate: how happy 1970's world would be to exchange population problems with 1930's Britain! Haldane, by contrast, only wrote a set of lectures, with a mathematical appendix; the lectures are outdated in many ways, but they are still well worth reading, and the appendix is full of surprises. Sewall Wright wrote a professional report which would have sufficed by itself to found population genetics; as in Fisher's case, but to a lesser degree, his originality as a mathematician makes him difficult to read.

Excellent summaries of population genetics for the beginner are (in increasing order of length), J. B. S. Haldane, "Natural Selection," pp. 101-149 in P. R. Bell, ed., *Darwin's Biological Work*, Cambridge University Press, 1959; the relevant chapters of Ehrlich and Holm's *The Process of Evolution*, McGraw-Hill, 1963; and D. S. Falconer, *Introduction to Quantitative Genetics*, Ronald Press, 1960. For proper perspective, one should also read the introduction to E. Adelberg, ed., *Papers on Bacterial Genetics*, Little, Brown, 1960: this tells of mutational equilibria, discusses chemostats as tools for experimenting with evolution, and brings much else to light that field biologists pass over. Mathematicians will find a brief and authoritative account of population genetics in G. Malecot, *Les Mathématiques de l'Hérédité,* Masson et Cie., 1948, recently reprinted in translation by W. H. Freeman as *The Mathematics of Heredity*, 1969. Very recently, two treatises of more comprehensive scope have appeared. J. Crow and M. Kimura, *Introduction to Population Genetics Theory*, Harper and Row, 1970, is a one-volume work which treats artificial as well as natural selection, an approach as useful today as it was in Darwin's time; most of this book should be accessible to my readers. Perhaps more difficult is S. Wright's summary work, *Evolution and the Genetics of Populations*, being put out by the University of Chicago Press, of which Volume I, *Genetic and Biometric Foundations* (1968), and Volume II, *The Theory of Gene Frequencies* (1969), have already appeared.

In their paper, "Non-Darwinian Evolution" (pp. 788-798 of *Science*, vol. 164, 1969), King and Jukes popularized the idea that most mutants are neutral, pointing out that replacements seem to occur at random in the chromosomes. However, this may only represent the random distribution of the sites where selection causes substitutions. Kimura's "Evolutionary Rate at the Molecular Level" (pp. 624-626 of *Nature*, vol. 217, 1968) emphasizes the striking similarity of the substitution rates in different evolving lines, and stresses the large number of these substitutions: he believes selection could not effect so many substitutions without destroying the population. Prakash and Lewontin, however, point out that even though most populations are extremely variable at the molecular level, so that each individual is heterozygous at a tenth of his loci, this variability is maintained by selection (S. Prakash and R. C. Lewontin, "A Molecular Approach to the Study of Genic Heterozygosity in Natural Populations III: Direct Evidence of Coadaptation in Gene Arrangements of *Drosophila*," pp. 398-405 of the *Proceed-*

ings of the National Academy of Sciences vol. 59, 1968). If this is true, how can most mutants be neutral? The problem will be a long time resolving.

For the frogs, see the opening chapter of E. Volpe, *Understanding Evolution*, W. C. Brown, 1967.

E. B. Ford analyzes several examples of selection in natural populations in his book, *Ecological Genetics*, Methuen, 1964.

CHAPTER 15

The Parliament of Genes

NATURAL selection favors the most reproductive genotype, even though it may harm the species. Thus selection favors the elaborate ornaments attracting a peacock's mate even though they make peacocks easier for predators to see. Some alleles spread by more blatant selfishness: a "segregation-distorter" allele spreads by somehow biasing (distorting) meiosis in its favor, so that most of the gametes, far more than the usual 50%, of an individual heterozygous for the distorter carry it rather than the normal allele.

Indeed, selection may even favor a gene which extinguishes its population. Consider a locus with alleles A and a in a population of haploids which live a year and mate only once, on Midsummer's Day. Suppose A is a "distorter": each mating produces two offspring, but the ratio of A to a among the offspring of A x a matings is $1 + e$: $1 - e$. Suppose, finally, that A also causes a heritable disease, killing a fraction s of its bearers, while all the a-bearers survive to mate. If the distortion effect is strong enough that the number of a-genes lost through mating with A

always exceeds the number of A-genes killed by disease, A will replace a, although the population declines to extinction (by a factor $1 - s$ per generation) as a result. For the mathematics of this process, let q_t be A's frequency among the mature individuals of generation t. Its frequency among the newborn of the next generation is the frequency of $A \times a$ matings, multiplied by the proportion $\frac{1}{2}(1 + e)$ of their offspring which carry A, plus the frequency of $A \times A$ matings: this sum is

$$q_t(1 - q_t)(1 + e) + q_t^2 = q_t + eq_t(1 - q_t).$$

This change is nearly equivalent to multiplying the ratio of A's frequency to a's by $1 + e$.* Selection during growth multiplies this ratio by $1 - s$. Distortion and disease together multiply the ratio of A's frequency to a's by $(1 + e)(1 - s)$ per generation: A spreads if this product exceeds 1.

How is the individual's advantage reconciled with the good of the group? How can a population defend itself from selfish genes? Lewontin describes a population of laboratory mice where he believes "group selection" arrested the spread of a distorter allele. His population was composed of many isolated, self-perpetuating groups of perhaps two males and six females each. 85% of the sperm of a male heterozygous for the distorter inherit this allele, but the males homozygous for it are sterile. The distorter spreads rapidly in groups it infects, but its spread

---

* The ratio of A's frequency to a's among the newborn offspring of generation t is

$$\frac{q_t + eq_t(1 - q_t)}{1 - q_t - eq_t(1 - q_t)} = \frac{q_t}{1 - q_t}\left\{\frac{1 + e(1 - q_t)}{1 - eq_t}\right\}.$$

Notice that

$$\left[\frac{1 + e(1 - q_t)}{1 - eq_t}\right]\left[\frac{1 + eq_t}{1 + eq_t}\right] = \frac{1 + e + e2q_t(1 - q_t)}{1 - e^2q_t^2}.$$

If e is so small that e^2 differs negligibly from 0, this factor is approximately $1 + e$. A more accurate approximation is $1 + e(1 + eq_t)$.

is arrested by the rapid extinction of infected groups. Put another way, the selective advantage of the distorter within infected groups is overcome by selection against infected groups. This is, however, the only case we know where selection between groups has overcome selection within the groups. Why is group selection so rarely effective?

If each infected group infects more than one other group before it dies out, the distorter spreads by a chain reaction: group selection is only effective if, of all the individuals infected as the distorter extinguishes a group, no more than one escapes to infect another group. Consider a group of N individuals, which we may imagine to represent N lines of descent. If each line of descent dies out k generations after it is infected, Nk individuals will be infected in all before the group dies. Thus group selection works only if less than one out of every Nk individuals migrates to another group. It is most effective against very harmful alleles which rapidly extinguish infected groups (making for small k)*, and it works most often in species composed of small, isolated groups. Such a species is better protected against disease of any sort: the history of the later Roman empire, or of Russia after her revolution, warns us how susceptible large, well-mixed populations can be to epidemics. Moreover, the small groups will evolve different genotypes, simply by chance: few diseases could wipe out all of them. However, evolution by sampling error has less desirable consequences: few species are composed of groups so small that selection amongst them can reconcile the individual's advantage with the good of the species.

---

* Consider a population of N individuals, all possessing the distorter allele. If this population declines by $1 - s$ each generation, the total number of individuals before it dies will be

$$N + N(1 - s) + N(1 - s)^2 + N(1 - s)^3 \ldots = N/s.$$

Thus if the distorter's disease has selective disadvantage s, it takes the distorter $1/s$ generations to kill an infected line of descent.

How do most populations protect themselves against selfish genes? If the distorter does not affect all its "normal" alleles, the resistant alleles will replace the sensitive, thus eliminating the distortion: the distorter then disappears if it is otherwise harmful. Such "distortion-resistant" alleles have evolved in yellow fever mosquitoes.

Moreover, a distorter allele only biases the segregation ratio for its own chromosome: its only effect on the genes of other chromosomes is to increase their death rate. Consequently, on these other chromosomes selection favors any allele suppressing all the effects of the distorter and thereby sparing its infected bearers the disease the distorter carries.

A distorter's spread may also be arrested by an allele on another chromosome which suppresses its distortion effect without suppressing the disease it carries. By assuring an honest meiosis, this allele spares some of its bearers' offspring a disease that would otherwise afflict them. To calculate the selective advantage of such modifiers, consider a distorter locus with alleles A and a, and a modifier locus on a different chromosome with alleles B and b, in a population of sexual haploids. Assume that if Ab mates with ab, the ratio of A to a among their offspring is $1 + e:1 - e$, whereas if one or both parents carry B, the ratio is $1 + \frac{1}{2}e:1 - \frac{1}{2}e$ or $1:1$, respectively. A B-gene thus halves the distortion among its bearers' offsping. In B's absence, A's frequency among newly conceived offspring exceeds that among their parents by $eq(1 - q)$, where q is A's frequency among the parents, so B-genes save a fraction $\frac{1}{2}eq(1 - q)$ of their bearers' offspring from the distorter. If the distorter kills a fraction s of its bearers, the selective advantage of the modifier is $\frac{1}{2}esq(1 - q)$. If the modifier is established (that is to say, if it is so numerous that it is unlikely to die out by chance) before the distorter has spread through all the population, it will insure the distorter's extinction: the distorter cannot persist when selection judges it only by its disease.

To the extent that different distorters bias meiosis by similar means, modifiers suppressing one allele's distortion are likely to be effective against another's. As the population copes with successive distorters, we might expect the transmission rules of meiosis to become ever more foolproof. It is as if we had to do with a parliament of genes: each acts in its own self-interest, but if its acts hurt the others, they will combine together to suppress it. The transmission rules of meiosis evolve as increasingly inviolable rules of fair play, a constitution designed to protect the parliament against the harmful acts of one or a few. However, at loci so closely linked to a distorter that the benefits of "riding its coattails" outweigh the damage of its disease, selection tends to enhance the distortion effect. Thus a species must have many chromosomes if, when a distorter arises, selection at most loci is to favor its suppression. Just as too small a parliament may be perverted by the cabals of a few, a species with only one, tightly linked chromosome is an easy prey to distorters.

Is there similar selection against subtler forms of selfishness? The striking beauty of many of the colors and forms designed to attract mates (like those designed to attract pollinating insects) suggests that our aesthetic sense has roots deep in our biological ancestry; it also suggests—how strange the paradox!—that selection often fails to control selfish mating practices. Perhaps the ornaments aren't costly: after all, should male display become dangerously extravagant, selection would favor those females mating with the less baroque males. When females have little choice in their mates, a more combative selfishness may evolve: some feel the Pleistocene Irish elk died out because its heavy, elaborate antlers were too clumsy. But its extinction suggests that nature does not tolerate unlimited deviations from "fair play."

Perfectly fair play insures that selection judges a genotype by the way its bearers cope with their environment, rather than how effectively they interrupt the survival or reproduction of other members of the species. It insures that individuals of different genotype compete as do members of different species: the genotype which spreads makes its species a better competitor. Thus the stronger the selection for fair play, the more closely is the individual's selective advantage identified with the good of the species.

Problem

The descendants of the Mayans apparently mistreat girls so badly that only one is raised for every two boys: the increased death rate caused thereby may be responsible for the mysterious disappearance of the Mayan civilization some 1500 years ago. Studies suggest (although they hardly prove) that the ratio of male births to female among the modern descendants of the Maya is 95:100, whereas among their neighbors it is 105:100 (data from Cowgill).

a) Could the tendency to mistreat girls be a genetical trait? If so, on what chromosome would the gene be? Justify your answer.

b) How strong a selection favors an allele shifting the Mayan sex ratio (at birth) from 102.5:97.5 to 90:110? Assume the ratio of male to female births in families with *AA* fathers, *Aa* fathers, and *aa* fathers is 102.5:97.5, 96.3:103.7, and 90:110, respectively, while the ratio of effort spent on male and female offspring is 133:67, 125:75, and 117:83, respectively. Assume the number of children raised is the same in every case.

c) Suppose that the Mayan sex ratio has changed because of an increase in *a*'s frequency, and that Mayans began to mistreat their girls 100 generations ago. Could *a* have been a recurrent mutant at that time?

Solutions:

a) This behavior hurts the population, but it causes a "distortion" in favor of males, suggesting that it could be caused by a gene on the Y-chromosome. Suppose there are two kinds of Y-chromosome, Y_1 and Y_2: a mating between XX and XY_1 leads to the successful raising of a man and a woman, while a mating between XX and XY_2 yields $1 + k$ men and $1 - k - e$ women. If women mate at random, and if a proportion p_t of the Y-chromosomes of generation t are Y_2, the proportion of Y_2 at $t + 1$ will be $p_{t+1} = p_t(1 + k)/(1 + kp_t)$.

b) If a's frequency is q_t in generation t, it will be $q_{t+1} = (1 - K_a\bar{K})q_t/(1 - \bar{K}^2)$ in generation $t + 1$, where $1 + K_a : 1 - K_a$ is the ratio of the effort spent raising male and female a-bearers, and $1 + \bar{K}:1 - \bar{K}$ is the ratio of the effort spent raising males and females in the population as a whole. q_{t+1} is very nearly $q_t + \bar{K}(\bar{K} - K_a)$. If a's frequency among the males is q_m, then $K_a = .17q_m + .25(1 - q_m)$ and $\bar{K} = .17q_m^2 + 2q_m(1 - q_m).25 + .33(1 - q_m)^2$, or $.17q_m + .33(1 - q_m)$. If q_t is nearly equal to q_m, then

$$q_{t+1} - q_t = q_t\bar{K}(\bar{K} - K_a) \approx .25q_t\,[.08(1 - q_m)],$$

which is very nearly $.02q_t(1 - q_t)$. Thus a's selective advantage is about 2%.

c) Before girls were mistreated, the sex ration 1.025:.975 divided effort equally among the sexes, while the sex ratio .90:1.10 divided effort in the ratio .875:1.125. Each generation, selection would diminish a's frequency an amount $\bar{K}(\bar{K} - K_a)q_t$, which is $.125q_t^2\,(.0625)(1 - q_t)$, or very nearly $.0078q_t^2(1 - q_t)$. At equilibrium, mutation (whose rate we call u) balances selection, so $u(1 - q) - .0078q^2(1 - q) = 0$, and $q = \sqrt{u/.0078}$. If the mutation rate u is 10^{-4}, q is very nearly 1/9 and $q/(1 - q)$ is nearly

1/8. When the mistreatment becomes habitual, dq/dt is $.02q(1 - q)$: the time required for this selection to shift the gene ratio from 1/8 to 2/3 is given by the equation

$$(1.02)^t(1/8) = 2/3.$$

84 generations are required.

Bibliographical Notes

Distortion effects are ably reviewed in S. Zimmering, L. Sandler, and B. Nicoletti, "Mechanisms of Meiotic Drive," pp. 409-436 of the *Annual Review of Genetics*, vol. 4, 1970. Their bibliography is copious and excellent.

R. H. MacArthur argues that selection will favor modifiers on other chromosomes suppressing a distortion of sex ratio: see "Population Effects of Natural Selection," pp. 195-199 of the *American Naturalist*, vol. 95, 1961. He feels this can cause autosomal control of sex ratio. Hiraizumi, Sandler, and Crow proposed that distortion would be suppressed by resistant alleles at the distorter locus: see "Meiotic Drive in Natural Populations of *Drosophila* III: Populational Implications of the Segregation-Distorter Locus," pp. 433-444 of *Evolution*, vol. 14, 1960. This is sometimes sufficient defense: mosquito sex is still controlled by a single locus.

W. A. Hickey and G. B. Craig, Jr., describe resistance to distortion by alleles at the same locus in "Distortion of Sex Ratio in Populations of *Aedes aegypti*," pp. 260-278 in the *Canadian Journal of Genetics and Cytology*, vol. 8, 1966. A gene modifying a distorter on a different chromosome is described in L. Sandler and A. Rosenfeld, "A Genetically Induced, Heritable Modification of Segregation-Distortion in *Drosophila melanogaster*," pp. 453-457 in the *Canadian*

Journal of Genetics and Cytology, vol. 4, 1962. Lewontin discusses group selection against a distorter in "Interdeme Selection Controlling a Polymorphism in the House Mouse," pp. 65-78 in the *American Naturalist*, vol. 94, 1962.

A popular history of epidemics is Hans Zinnser's *Rats, Lice and History*, Little, Brown, 1935, reprinted in 1960 by Bantam Paperbacks.

The extent to which selection favors the good of the group is the subject of a violent controversy. Many naturalists know of adaptations which apparently favor the good of the group over the individual's advantage, and willingly accept group selection as an explanation: geneticists know perfectly well group selection doesn't work, but often feel compelled to strengthen their case by denying the validity of their opponents' observations. See, for example, V. C. Wynne-Edwards, *Animal Dispersion in Relation to Social Behavior*, Oliver and Boyd, 1962, and George Williams, *Adaptation and Natural Selection*, Princeton University Press, 1966. Those wishing briefer summaries of the controversy should consult J. A. Wiens, "On Group Selection and Wynne-Edwards' Hypothesis," pp. 273-287 of the *American Scientist*, vol. 54, 1966; A. F. Skutch, "Adaptive Limitation of the Reproductive Rate of Birds," pp. 579-599 of *Ibis*, vol. 109, 1967; and V. C. Wynne-Edwards, "Intergroup Selection in the Evolution of Social Behavior," pp. 623-626 of *Nature*, vol. 200, 1963.

CHAPTER 16

The Individual's Advantage and the Good of the Group

HOW DOES an allele's selective advantage relate to its contribution to the good of the species? To answer, we must define our terms more carefully. Consider a locus with alleles A and a in a population whose generations overlap (a population where, as among humans, individuals are continually dying and being born). Suppose, to start with, that selection acts only at this locus.

The difference between the contributions to fitness of A- and a-genes, which for short we call their difference in fitness, is the difference in fitness between a population composed solely of A-bearers and one of otherwise identical genetics composed solely of a-bearers. In a space-limited population where A-bearers can tolerate K_A individuals per acre in their surroundings, while a-bearers can tolerate only K_a, the difference in fitness between A and a is $K_A - K_a$. In a food-limited population where A-bearers can survive when there are R_A prey animals per acre, while a-bearers need R_a, the difference in fitness between A and a is $R_a - R_A$ (the genotype needing less food is obviously more fit).

Were successive generations distinct, we would define the selective advantage s of A over a as the proportion by which the "gene ratio," the ratio of A's numbers to a's, changes in one generation. If each generation the numbers of A- and a-bearers multiply by W_A and W_a, respectively, s is given by

$$\frac{W_A}{W_a} = 1 + s\,;$$

$$s = \frac{W_A - W_a}{W_a}\,.$$

When generations overlap, we do not know how long a generation is; it is easier to measure the change in gene ratio per day or year. We then define the selective advantage of A over a as the proportional rate of change in the gene ratio: the rate of change divided by the gene ratio itself. In symbols,

$$s = \frac{N_a}{N_A}\left(\frac{d}{dt}\frac{N_A}{N_a}\right) = \frac{1}{N_A}\frac{dN_A}{dt} - \frac{1}{N_a}\frac{dN_a}{dt},$$

where s is A's selective advantage, and $N_A(t)$ and $N_a(t)$ are the numbers of A- and a-bearers at time t. The selective advantage of A over a is thus the difference in their rates of multiplication. We correspondingly define the selective advantage of one genotype over another as the difference in their multiplication rates, were there no recombination among their constituent genes.

If A and a differ only slightly in fitness, A's selective advantage will be nearly constant, and proportional to the difference in fitness: A's gradual spread does not alter the environment enough to change its selective advantage. For example, in a space-limited population the replacement of a by A is described (Chapter 7) by the equation

$$\frac{d}{dt}\log\frac{N_A}{N_a^x} = \frac{1}{N_A}\frac{dN_A}{dt} - \frac{x}{N_a}\frac{dN_a}{dt} = a_A(K_A - K_a),$$

where N_A and N_a are the populations of A and a, K_A and K_a their crowding tolerances, a_A and a_a describe how fast bearers of the two alleles crowd their environments, and $x = a_A/a_a$. If x is close to 1, the gene ratio N_A/N_a changes by a nearly geometric progression, and the selective advantage of A over a is nearly $a_A(K_A - K_a)$. Similar argument shows that in food-limited species also, a constant fitness difference implies a nearly constant selective advantage.

How fast can selection at this locus improve the population's fitness? The effectiveness of selection is governed by the amount of variability available for it to act upon. The fitness of a space-limited population is the fitness of A-bearers, times the proportion of the population carrying A, plus the corresponding product for a-bearers. In symbols, the fitness $\bar{K}$ of the population is $qK_A + (1 - q)K_a$, where q is A's frequency. The rate of increase of population fitness is the increase from replacing a by A, times the rate at which A replaces a:

$$\frac{d\bar{K}}{dt} = (K_A - K_a)\frac{dq}{dt}.$$

The preceding paragraph tells us that

$$\frac{d}{dt}\log\frac{q}{1-q} = \frac{d}{dt}\log\frac{N_A}{N_a} \approx a_A(K_A - K_a).$$

We therefore conclude that

$$\frac{dq}{dt} = a_A(K_A - K_a)\,q(1-q);$$

$$\frac{d\bar{K}}{dt} = a_A(K_A - K_a)^2\,q(1-q).$$

How does the speed of fitness increase relate to the variability available for selection? We may measure this variability by the degree to which the fitnesses of A and a differ from that of the population as a whole. The simplest measure is the variance in allele fitnesses, usually called the genic variance in fitness: in symbols, this is

$$q(K_A - \bar{K})^2 + (1-q)(K_a - \bar{K})^2.$$

Substituting $qK_A + (1-q)K_a$ for $\bar{K}$, this expression reduces to

$$q(1-q)(K_A - K_a)^2.$$

The effectiveness of selection, as measured by how fast it increases the population's fitness, is thus proportional to the genic variance in fitness. This is equally true for food-limited species, etc.

Suppose now that selection acts simultaneously on two loci, with alleles A, a and B, b, respectively. Here, A may spread either through its own contribution to fitness or from association with the allele favored at the other locus. How are we to distinguish the contributions to fitness of A and B? Does genetic variance still measure the variability available for selection?

Consider a haploid, space-limited population. Suppose that AB, Ab, aB, and ab can tolerate K_{AB}, K_{Ab}, K_{aB}, and K_{ab} individuals per acre in their environment; and let the frequencies of these genotypes be x_{AB}, x_{Ab}, x_{aB}, and x_{ab}, respectively. The frequency p of A* is then $x_{AB} + x_{Ab}$; the frequency q of B is $x_{AB} + x_{aB}$. The selective advantage of A over a is proportional to the difference s_A between the average crowding tolerances of A- and a-bearers: in symbols,

$$s_A = \frac{K_{AB}\,x_{AB} + K_{Ab}\,x_{Ab}}{x_{AB} + x_{Ab}} - \frac{K_{aB}\,x_{aB} + K_{ab}\,x_{ab}}{x_{aB} + x_{ab}}.$$

In future, we will speak of s_A as if it were the selective advantage of A over a. Similarly,

* Notice the change from preceding paragraphs, where q denoted A's frequency.

$$s_B = \frac{K_{AB}\,x_{AB} + K_{aB}\,x_{aB}}{x_{AB} + x_{aB}} - \frac{K_{Ab}\,x_{Ab} + K_{ab}\,x_{ab}}{x_{Ab} + x_{ab}}.$$

The difference in fitness between two alleles is simply the effect on fitness of replacing one by the other. Let α be the effect of replacing a by A, and β that of replacing b by B. Were the loci acting independently, the crowding tolerances of ab, aB, Ab, and AB would be K_{ab}, $K_{ab}+\beta$, $K_{ab}+\alpha$, and $K_{ab}+\alpha+\beta$, respectively: in general, we try to fit such an additive model as best we can to the observed facts. To this end, we seek K,α, and β minimizing the sum

$$x_{AB}(K_{AB} - K - \alpha - \beta)^2 + x_{Ab}(K_{Ab} - K - \alpha)^2$$
$$+ x_{aB}(K_{aB} - K - \beta)^2 + x_{ab}(K_{ab} - K)^2.$$

We are in effect minimizing the variance of the actual fitnesses from the additive model. Selective advantages and fitness differences are related as follows:*

---

* Let Q be the variance of the fitnesses from the model. We find K,α, and β from the equations

1) $\dfrac{\delta Q}{\delta K} = 0 = -2(K_{AB}\,x_{AB} + K_{Ab}\,x_{Ab} + K_{aB}\,x_{aB} + K_{ab}\,x_{ab} - K - p\alpha - q\beta).$

2) $\dfrac{\delta Q}{\delta \alpha} = 0 = 2[p(K+\alpha) + x_{AB}\beta - K_{AB}\,x_{AB} - K_{Ab}\,x_{Ab}].$

3) $\dfrac{\delta Q}{\delta \beta} = 0 = 2[q(K+\beta) + x_{AB}\alpha - K_{AB}x_{AB} - K_{aB}\,x_{aB}].$

Subtracting p times the first equation from the second, we obtain

$$p(1-p)\alpha + (x_{AB} - pq)\beta$$
$$-[(1-p)(K_{AB}\,x_{AB} + K_{Ab}\,x_{Ab}) - p(K_{aB}\,x_{aB} + K_{ab}\,x_{ab})] = 0.$$

This equation may be expressed more simply as

$$p(1-p)\alpha + (x_{AB} - pq)\beta = p(1-p)s_A.$$

Subtracting q times the first equation from the third, we obtain the corresponding equation for s_B.

$$p(1-p)s_A = p(1-p)\alpha + (x_{AB} - pq)\beta;$$

$$q(1-q)s_B = q(1-q)\beta + (x_{AB} - pq)\alpha.$$

If selection favors B, A derives a selective advantage from association with B: to be specific, A spreads if x_{AB} exceeds pq. An allele's selective advantage is identical to its contribution to fitness only if the alleles of different loci associate at random, so that the frequency of each genotype is the product of the frequencies of its component alleles.

How rapidly does selection improve fitness? The population's variability is best measured by the genic variance in the fitnesses predicted by the linear model for the different genotypes. This variance is

$$x_{AB}(K + \alpha + \beta - \bar{K})^2 + x_{Ab}(K + \alpha - \bar{K})^2$$

$$+ x_{aB}(K + \beta - \bar{K})^2 + x_{ab}(K - \bar{K})^2 .$$

This may be expressed more simply as*

$$\alpha s_A p(1-p) + \beta s_B q(1-q).$$

---

* To evaluate the variance, set $\bar{K} = K + \alpha p + \beta q$. We may express our sum as

$$x_{AB}[(1-p)\alpha + (1-q)\beta]^2 + x_{Ab}[(1-p)\alpha - q\beta]^2$$

$$+ x_{aB}[(1-q)\beta - p\alpha]^2 + x_{ab}[-p\alpha - q\beta]^2.$$

This simplifies to

$$p(1-p)\alpha^2 + q(1-q)\beta^2 + 2\alpha\beta[x_{AB}(1-p)(1-q)$$

$$- x_{Ab}(1-p)q - x_{aB}p(1-q) + x_{ab}pq].$$

Notice that if $d = x_{AB} - pq$, then $x_{Ab} = p - pq - d = p(1-q) - d$; $x_{aB} = q(1-p) - d$; and $x_{ab} = (1-p)(1-q) + d$. Substituting these expressions, we find the coefficient of $2\alpha\beta$ is simply $x_{AB} - pq$. The genic variance may therefore be expressed as

$$\alpha[p(1-p)\alpha + \beta(x_{AB} - pq)] + \beta[q(1-q)\beta + \alpha(x_{AB} - pq)]$$

$$= p(1-p)\alpha s_A + q(1-q)\beta s_B.$$

Since $dp/dt = \tilde{a}s_A p(1-p)$ and $dq/dt = \tilde{a}s_B q(1-q)$, where $\tilde{a}$ is a constant measuring how fast the population crowds its environment, the genic variance in fitness is proportional to $\alpha dp/dt + \beta dq/dt$. Is this not the rate of fitness increase, the increase from replacing a by A, times the speed of this replacement, plus the corresponding product for B?

In fact, genic variance measures selection's effectiveness only if the "linkage cross-ratio" $x_{AB}x_{ab}/x_{aB}x_{Ab}$ is constant: when this ratio changes, fitness increase cannot be expressed as $\alpha dp + \beta dq$. (See Appendix.) This cross-ratio is unity if alleles at different loci associate at random. If the proportion of the population undergoing recombination between the two loci each generation greatly exceeds the proportion by which the gene ratio at either locus changes, the linkage cross-ratio rapidly approaches a steady-state value, and genic variance measures selection's effectiveness ever more accurately. Alleles are associated at random in this steady state only if the fitness contributions of A and B are additive; that is to say, only if $K_{AB} - K_{ab} = K_{Ab} - K_{ab} + K_{aB} - K_{ab}$, or if $K_{AB} - K_{Ab} - K_{aB} + K_{ab} = 0$. The greater the departure from additivity, and the stronger the linkage between the two loci, the more the steady-state cross-ratio differs from 1 (see Appendix).

Appendix – Proof of Fisher's "Fundamental Theorem of Natural Selection"

1) When can $d\bar{K}$ be expressed as $\alpha dp + \beta dq$?

If genotypic fitnesses are constant, $d\bar{K} = K_{AB}\,dx_{AB} + K_{Ab}\,dx_{Ab} + K_{aB}\,dx_{aB} + K_{ab}\,dx_{ab}$. Let $\theta_{AB} = x_{AB}/pq$, $\theta_{Ab} = x_{Ab}/p(1-q)$, etc: these constants are all unity if alleles associate at random. Writing $x_{AB} = pq\theta_{AB}$, etc., we may express $d\bar{K}$ as the following sum:

$$K_{AB}\,[p\theta_{AB}\,dq + q\theta_{AB}\,dp + pqd\theta_{AB}]$$
$$+ K_{Ab}\,[-p\theta_{Ab}\,dq + (1-q)\theta_{Ab}\,dp + p(1-q)d\theta_{Ab}]$$
$$+ K_{aB}\,[(1-p)\theta_{aB}\,dq - q\theta_{aB}\,dp + (1-p)qd\theta_{aB}]$$
$$+ K_{ab}\,[-(1-p)\theta_{ab}\,dq - (1-q)\theta_{ab}\,dp + (1-p)(1-q)d\theta_{ab}].$$

The sum of the terms multiplying dq is

$$p\theta_{AB}K_{AB} + (1-p)\theta_{aB}K_{aB} - p\theta_{Ab}K_{Ab} - (1-p)\theta_{ab}K_{ab}.$$

Recalling that $p\theta_{AB} = x_{AB}/q$, etc., we may rewrite this as

$$\frac{1}{q}(x_{AB}K_{AB} + x_{aB}K_{aB}) - \frac{1}{1-q}(x_{Ab}K_{Ab} + x_{ab}K_{ab}),$$

which expression is equal to s_B. Similarly, dp is multiplied by s_A. Substituting $x_{AB}\,d\log\theta_{AB}$ for $pqd\theta_{AB}$, etc., we find

$$d\bar{K} = s_A\,dp + s_B\,dq + x_{AB}K_{AB}\,d\log\theta_{AB}$$
$$+ x_{Ab}K_{Ab}\,d\log\theta_{Ab} + x_{aB}K_{aB}\,d\log\theta_{aB} + x_{ab}K_{ab}\,d\log\theta_{ab}.$$

Now substitute

$$s_A = \frac{(x_{AB}-pq)\beta}{p(1-p)} + \alpha;$$

$$s_B = \beta + \frac{(x_{AB}-pq)\alpha}{q(1-q)}$$

to obtain

$$d\bar{K} = \alpha dp + \beta dq + \frac{(x_{AB}-pq)\beta dp}{p(1-p)} + \frac{(x_{AB}-pq)\alpha dq}{q(1-q)}$$

$+$ terms in $d\log\theta$.

To simplify $(x_{AB}-pq)\beta dp/p(1-p)$, notice that $q = x_{AB} + x_{aB}$, or $pq\theta_{AB} + (1-p)q\theta_{aB}$, so that $1 = p\theta_{AB} + (1-p)\theta_{aB}$.

Differentiating the latter and multiplying by q, we find

$$qpd\theta_{AB} + q(1-p)d\theta_{aB} = q(\theta_{aB} - \theta_{AB})dp$$

$$= [\frac{x_{aB}}{1-p} - \frac{x_{AB}}{p}]dp = \frac{dp}{p(1-p)}[px_{aB} - (1-p)x_{AB}].$$

Setting $x_{aB} = q - x_{AB}$, we may replace our expression by

$$\frac{dp}{1-p}[p(q - x_{AB}) - (1-p)x_{AB}],$$

or

$$\frac{dp}{p(1-p)}(pq - x_{AB}).$$

Writing $pqd\theta_{AB}$ as $x_{AB}d \log \theta_{AB}$, etc., we obtain

$$\frac{(pq - x_{AB})\beta dp}{p(1-p)} = \beta(x_{AB}d \log \theta_{AB} + x_{aB}d \log \theta_{aB}).$$

Similarly,

$$\frac{(pq - x_{AB})\alpha dq}{q(1-q)} = \alpha(x_{AB}d \log\theta_{AB} + x_{Ab}d \log\theta_{Ab}).$$

Incorporating all this information, we find

$$d\overline{K} = \alpha dp + \beta dq + x_{AB}(K_{AB} - \alpha - \beta)d \log \theta_{AB}$$

$$+ x_{Ab}(K_{Ab} - \alpha)d \log \theta_{Ab} + x_{aB}(K_{aB} - \beta)d \log \theta_{aB}$$

$$+ x_{ab}\, d \log \theta_{ab} K_{ab}.$$

Since $-K[pq\, d\theta_{AB} + (1-p)q\, d\theta_{aB} + p(1-q)d\theta_{Ab} + (1-p)(1-q)d\theta_{ab}]$ is zero*, we may write

* We may prove this by setting $x_{AB} = pq + D$, $x_{Ab} = p(1-q) - D$, etc. Then $\theta_{AB} = 1 + D/pq$ and $pq\, d\,\theta_{AB} = dD - D(dq/q + dp/p)$. Once we find similar expressions for $p(1-q)d\theta_{Ab}$, etc., it is very easy to see that their sum vanishes.

$$d\bar{K} = \alpha dp + \beta dq + x_{AB}\, d \log \theta_{AB} (K_{AB} - K - \alpha - \beta)$$
$$+ x_{Ab}\, d \log \theta_{Ab} (K_{Ab} - K - \alpha)$$
$$+ x_{aB}\, d \log \theta_{aB} (K_{aB} - K - \beta)$$
$$+ x_{ab}\, d \log \theta_{ab} (K_{ab} - K)$$
$$= \alpha dp + \beta dq + x_{AB} d_{AB}\, d \log \theta_{AB} + x_{Ab} d_{Ab} d \log \theta_{Ab} + \ldots,$$

where $d_{AB} = K_{AB} - K - \alpha - \beta$, $d_{Ab} = K_{Ab} - K - \alpha$, $d_{aB} = K_{aB} - K - \beta$, and $d_{ab} = K_{ab} - K$; these constants measure the deviation of the fitnesses of the different genotypes from those predicted by the additive model. K, α, and β were chosen to minimize

$$Q = x_{AB} d_{AB}^2 + x_{Ab} d_{Ab}^2 + x_{aB} d_{aB}^2 + x_{ab} d_{ab}^2.$$

The equations $\delta Q/\delta K = 0$, $\delta Q/\delta\alpha = 0$, and $\delta Q/\delta\beta = 0$ imply, respectively:

$$x_{AB} d_{AB} + x_{Ab} d_{Ab} + x_{aB} d_{aB} + x_{ab} d_{ab} = 0;$$
$$x_{AB} d_{AB} + x_{Ab} d_{Ab} = 0;$$
$$x_{AB} d_{AB} + x_{aB} d_{aB} = 0.$$

Therefore, $x_{AB} d_{AB} = -x_{Ab} d_{Ab} = -x_{aB} d_{aB} = x_{ab} d_{ab}$, so we may express $d\bar{K}$ as

$$\alpha dp + \beta dq + x_{AB} d_{AB} (d \log \theta_{AB} + d \log \theta_{ab}$$
$$- d \log \theta_{Ab} - d \log \theta_{aB}).$$

This last term vanishes either if the loci act additively ($d_{AB} = 0$) or if the linkage cross-ratio is constant

$$(d \log (x_{AB} x_{ab} / x_{Ab} x_{aB}) = d \log (\theta_{AB} \theta_{ab} / \theta_{Ab} \theta_{aB}) = 0).$$

2) When is the linkage cross-ratio constant?

Were AB, Ab, aB, and ab four alleles at a single locus, dx_{AB}/dt would be

$$\frac{d}{dt}\frac{N_{AB}}{N} = \frac{1}{N}\frac{dN_{AB}}{dt} - \frac{N_{AB}}{N}\left(\frac{1}{N}\frac{dN}{dt}\right),$$

where N_{AB} is the number of AB-bearers and N the total population size. Thus

$$\frac{dx_{AB}}{dt} = \tilde{a}(K_{AB} - \bar{K})x_{AB},$$

where $\tilde{a}$ relates fitness differences to selective advantages. If, however, a proportion k of the population undergoes recombination per unit time, dx_{AB}/dt will be diminished by an amount $kx_{AB}x_{ab}$ from recombination between AB and ab to form Ab and aB; it will be increased by $kx_{Ab}x_{aB}$ due to the formation of AB from Ab and aB. Thus

$$\frac{dx_{AB}}{dt} = \tilde{a}(K_{AB} - \bar{K})x_{AB} - k(x_{AB}x_{ab} - x_{Ab}x_{aB}).$$

Letting D represent the expression $x_{AB}x_{ab} - x_{Ab}x_{aB}$, we may write

$$\frac{dx_{AB}}{dt} = \tilde{a}(K_{AB} - \bar{K})x_{AB} - kD,$$

$$\frac{dx_{Ab}}{dt} = \tilde{a}(K_{Ab} - \bar{K})x_{Ab} + kD,$$

$$\frac{dx_{aB}}{dt} = \tilde{a}(K_{aB} - \bar{K})x_{aB} + kD,$$

$$\frac{dx_{ab}}{dt} = \tilde{a}(K_{ab} - \bar{K})x_{ab} - kD.$$

Let R be the linkage cross-ratio, $x_{AB}x_{ab}/x_{Ab}x_{aB}$. Then

$$\frac{d\log R}{dt}=\frac{d\log x_{AB}}{dt}-\frac{d\log x_{Ab}}{dt}-\frac{d\log x_{aB}}{dt}+\frac{d\log x_{ab}}{dt}$$

$$=\tilde{a}[K_{AB}-K_{Ab}-K_{aB}+K_{ab}]$$

$$-kD\left[\frac{1}{x_{AB}}+\frac{1}{x_{Ab}}+\frac{1}{x_{aB}}+\frac{1}{x_{ab}}\right].$$

Notice that

$$D=x_{Ab}x_{aB}(R-1),$$

$$\frac{1}{x_{AB}}+\frac{1}{x_{ab}}=\frac{x_{AB}+x_{ab}}{x_{ab}x_{AB}},$$

$$\frac{1}{x_{Ab}}+\frac{1}{x_{aB}}=\frac{x_{Ab}+x_{aB}}{x_{Ab}x_{aB}},$$

$$kD\left\{\frac{1}{x_{AB}}+\frac{1}{x_{aB}}+\frac{1}{x_{Ab}}+\frac{1}{x_{ab}}\right\}$$

$$=k(R-1)\left\{\frac{x_{AB}+x_{ab}}{R}+x_{Ab}+x_{aB}\right\}.$$

If R is close to 1, this is nearly $k(R-1)$, and

$$\frac{d\log R}{dt}=\tilde{a}[K_{AB}-K_{Ab}-K_{aB}+K_{ab}]-k(R-1).$$

Thus, if $\tilde{a}/k$ is much less than 1, R approaches the steady-state value R', where

$$R'=1+\frac{\tilde{a}}{k}[K_{AB}-K_{Ab}-K_{aB}+K_{ab}].$$

Bibliographical Notes

The relation between fitness increase and genetic variance is Fisher's Fundamental Theorem of Natural Selection, which he first proved in 1930. The clearest expositions of it are R. A. Fisher, "Average Excess and Average Effect of a Gene Substitution," pp. 53-63 of *Annals of Eugenics* (now *Annals of Human Genetics*), vol. 11, 1941; and M. Kimura, "On the Change of Population Fitness by Natural Selection," pp. 145-167 of *Heredity*, vol. 12, 1958.

Our proof is drawn from Fisher's 1941 paper and Kimura's "Attainment of Quasi-linkage Equilibrium when Gene Frequencies are Changing by Natural Selection," pp. 875-890 of *Genetics*, vol. 52, 1965. Fisher's distinction between the average excess and the average effect of a gene substitution corresponds to the distinction between an allele's selective advantage and its contribution to fitness. Fisher and Kimura measure fitness by rate of multiplication, which denies their theorem predictive value, for who ever heard of a population increasing indefinitely at an ever-increasing rate? (Some act as if humans can do this, but this seems as bad theology as it is biology: surely Luke 4: 9-12 and Matthew 4: 5-7 warn us not to expect miracles to save us from the otherwise certain consequences of our actions.) MacArthur solved the paradox by assigning fitness an ecological meaning: see his paper, "Some Generalised Theorems of Natural Selection," pp. 1893-1897 of the *Proceedings of the National Academy of Sciences*, vol. 48 (1962).

How useful is the Fundamental Theorem? A fundamental paper by Ian Franklin and R. C. Lewontin, "Is the Gene the Unit of Selection?" (pp. 707-734 of *Genetics*, vol. 65, 1970), suggests that conditions permitting "quasi-linkage equilibria" occur rather rarely.

Our interpretation of the theorem can only apply to a population occupying a distinct and clearly defined niche (otherwise we would lack an "ecological" measure of fitness). Even then we must assume that the population's size is near equilibrium, and that this equilibrium is not much disturbed by the gene substitutions in progress. This sort of theory may well describe the evolution of the species of a mature community, but it can hardly describe the evolution of populations invading new niches, whose limiting factors are not yet fully developed. This is a serious weakness; for the "race" to occupy new or newly emptied niches is one of the most important evolutionary processes.

CHAPTER 17

Coping with Change

HOW IS a population affected by environmental change? The environment changes according to many time scales. An allele may be favored one generation and selected against the next, or it may be favored for thousands of generations in succession. Some changes occur again and again, like dry years; some are slow and irreversible, like the drying out of the coal-swamp world of the Permian; some, such as glaciation, are sudden and seemingly unique.

Let us turn first to recurrent change: consider a locus with two alleles, *A* and *a*, in a population whose environment has two states, *I* and *II*. A completely free response to environmental change eventually deprives the population of its ability to respond: if environment *I* favors *A* and *II* favors *a*, their alternation will shift allele frequencies until one allele disappears and the population can no longer respond to selection. This is true even if, on the average, selection favors both alleles equally: suppose, for example, that a generation of environment *I* multiplies the ratio of *A*'s frequency to *a*'s by $1 + s$, while a generation of *II* multiplies it by $1/(1 + s)$. m generations of *I* and n of

II therefore multiply the gene ratio by $(1 + s)^{m-n}$. Even if *I* and *II* are equally likely, the vagaries of chance will bring a time when m differs enough from n that one allele disappears.

If the environmental states follow at random, selection during one generation cannot prepare the population for the environment of the next, and a fixed strategy is the best response. Among micro-organisms, different generations may experience day and night: a week's spell of rainy weather may span several generations. For such small and short-lived creatures, the environment changes drastically, and often unpredictably, from one generation to the next. Perhaps micro-organisms are so often asexual because it does not pay them to evolve, because it is self-defeating to try adjusting to the changes of an apparently random environment.

If, however, the environment remains the same for several generations, selection can adapt the population, and it would pay our ideal population to maintain both A and a. If a proportion u of the population's A-bearers are replaced each generation by mutant or migrant a-bearers and vice versa, neither allele can disappear. Allele frequencies respond the more readily to selection the nearer they are to 1/2. The higher the replacement rate, the more adaptable the population. On the other hand, a high replacement rate prevents complete adaptation, for the population will have many individuals adapted to the other environment.

Suppose for the moment that mutation or migration is the only answer to change. The optimum replacement rate balances adaptability and adaptedness, and is the higher for more changeable environments. To calculate this optimum, suppose the environment is stable long enough for the frequency of the favored allele to reach equilibrium. If A is favored, at equilibrium its frequency is nearly 1, and selection must eliminate a proportion u of the population each generation to compensate for the replacement of A by a. If a's selective disadvantage is s, selection eliminates a proportion s of the a-bearers each generation, so a's frequency at equilibrium is u/s.*

Suppose now that the environment changes and a acquires selective advantage s over A. What fraction of the population must be eliminated to substitute a for A? The total cost of the substitution is the sum of each generation's "selective deaths." Most of these individuals die before a's frequency attains 1/2. During most of this period, A's frequency is nearly 1, so the substitution's cost is A's selective disadvantage s times the number x of generations required to shift the gene ratio from u/s to 1. x is given by the equations

$$\frac{u}{s}(1+s)^x = 1; \qquad x \log(1+s) \cong xs = \log\frac{s}{u}.$$

A substitution every n generations reduces the population by a fraction $(1/n)\log(s/u)$, on the average.† The replacement rate u best balancing the cost of substitution against the cost of eliminating ill-adapted mutants or migrants is that which minimizes the proportion $f(u)$ of the population selection must eliminate, where

$$f(u) = u + \frac{1}{n}\log\frac{s}{u}.$$

The optimum u is $1/n$: we find it by setting $df/du = 0$.

---

* Selection against a eliminates a proportion of the population equal to a's frequency times the proportion of a-bearers eliminated. In symbols, $u = sq$, where q is a's equilibrium frequency.

† To calculate more precisely the cost of an allele substitution, notice that, when a becomes favored, selection eliminates a proportion sq of each generation, where q is A's frequency. The accumulated loss due to the time required to substitute a for A is

$$s\int q(t)\,dt = s\int_{1-u/s}^{u/s} q\frac{dt}{dq}\,dq = s\int_{1-u/s}^{u/s}\frac{q}{-sq(1-q)}\,dq$$

$$= -\int_{1-u/s}^{u/s}\frac{dq}{1-q} = -\log\frac{u}{s}.$$

How does a population prevent over-response to an environment which changes every ten or hundred generations? If these changes are local, immigrants preserve adaptability. Unfortunately, they hinder adaptation to any local peculiarity of the habitat: migration must balance the advantages of responding quickly to temporary changes and of specializing to different habitats. An opportunistic fruit fly, whose generations sense the change in weather from one fortnight to the next, relies on the dispersal powers of its migrants to prevent over-response to local change, but it pays an animal whose life spans all seasons to adapt more fully to its chosen habitat. For like reasons, the vegetation zones of a mountainside are defined by long-lived trees, while the weeds and flowers of the understory span several such zones. Were responsiveness perfectly adjusted, the trees of a kind would flower in a wave advancing up the mountain, while flowers of a kind would bloom all at once to exchange genes from far away: however, the glacier lilies and marsh marigolds following the snow's edge testify conflicting needs.

Balanced polymorphisms also regulate response to change. The English land snail, *Cepaea nemoralis*, must match its background to avoid being eaten by thrushes. Snails living in grass or weeds are usually green, while those living among the dead leaves of a woodland floor are usually pink or brown. Moreover, snails from uniform backgrounds have unbanded shells, but the shells of those whose background is the variegated pattern of lawn or hedge are banded. Color is determined by a single locus: the heterozygotes resemble the pink or brown homozygotes, but they possess some physiological advantage which maintains both green and brown snails in most populations. Bandedness is controlled by other, closely linked loci. The polymorphism is necessary partly because the color matching the many brown backgrounds of early spring is ill suited to late May's foliage, but it also helps the snails adapt to slower changes, the neglected field growing into woodland which is then cut over and left to weeds.

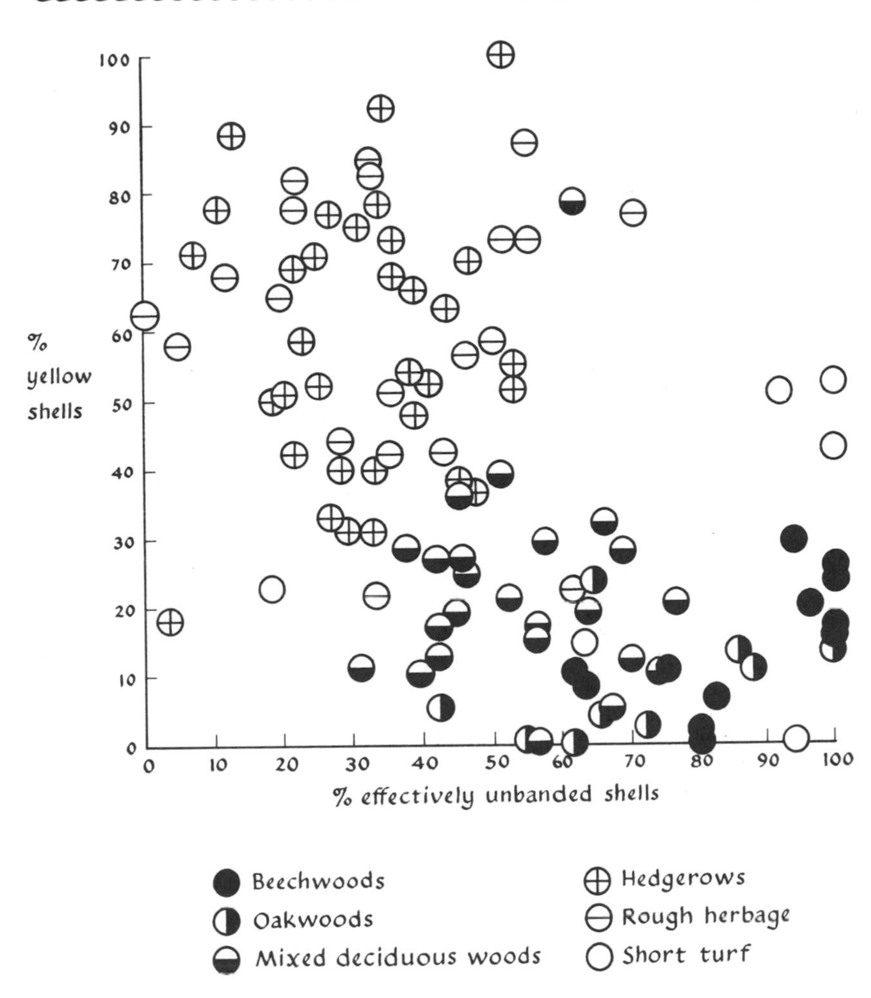

Fig. 17-1. Each circle represents a population of snails: the position of the circle denotes the proportions of yellow and of unbanded snails in the population, and the design within the circle indicates the nature of the population's habitat. Notice that uniformly pink and brown snails prevail in beechwoods, while banded yellow ones prevail in hedgerows and rough herbage. (After Fig. 1 of A. J. Cain and P. M. Sheppard, "Natural Selection in *Cepaea*," p. 99 of *Genetics*, vol. 39, 1954, with permission of the Genetics Society of America.)

The seasons bring many changes at once. Winter exposes fruit flies to cold, but releases them from overcrowding. If one locus controlled temperature and another crowding tolerance, it would pay the population to link these into a single "supergene": summer's advent would then cost one allele substitution, a heat-loving, crowd-tolerant allele for a cold-tolerant one, rather than two. In general, should the selective values of several characters fluctuate simultaneously, these should be controlled by a single gene. Many balanced polymorphisms involve supergenes: perhaps the most striking example is *Drosophila pseudoobscura*, whose chromosomes may exhibit any of several gene arrangements, between which there is no cross-over. An individual possessing homologous chromosomes with different arrangements is usually more fit than a "homozygote." Frequencies of these arrangements fluctuate with the seasons, and the selective influences involved can often be verified in the laboratory.

The changes affecting animals longer lived than flies or snails are slower, and affect one locus at a time. Here, linkage should be so loose that a substitution at one locus does not affect others. Genes at different loci should act independently of each other: their effects should be additive. Even though genes interact in very complex ways, selection can produce additivity. In a curious paper, Crow compares the genetics of poison resistance in fruit flies and bacteria. The genes contributing to the chloramphenicol resistance of his bacteria are selected for fitness in the genetic backgrounds where they first appear, for there was no opportunity for recombination. The effects of these genes are by no means additive. On the other hand, the genes contributing to DDT resistance in fruit flies must function in many backgrounds, so alleles are selected which contribute the same to fitness whatever the composition of the rest of the genome.*

* The genome denotes the ensemble of genes an organism possesses.

If the environment remains stable for thousands or millions of generations, it pays the population to respond quite freely to a change: the ordinary mutability of genes insures sufficient responsiveness. Mutations are therefore a necessary adaptation, not merely harmful accidents which selection would prevent if it only could. Mutation rates are not as low as possible: in bacteria, one can select for mutability a hundredfold less than normal. Are mutation rates in fact suited to environmental change, or are they no lower because exact replication costs more than it is worth? To decide, we must know how mutations arise, and how they can be prevented.

Mutants occur either as changes of state in resting DNA molecules or as replication errors: in slowly reproducing bacterial cultures, mutants occur proportionally to elapsed time, as if they occurred at random in resting molecules, whereas in rapidly reproducing cultures the number of mutants is more nearly proportional to the number of replications. Bacteria seem to correct most replication errors: an enzyme apparently goes along a newly formed DNA strand, while it is still paired to its parent, snipping out improperly paired sections so they can form again. Greater replicative accuracy probably requires more enzymes, the protein for which finds better use elsewhere. We also know that some nucleotides (the so-called "hot spots" of the genome) are more mutable than others: a nucleotide's mutability must be affected by the arrangement of its neighbors. An allele's mutability could probably be lowered, without cost and perhaps without changing the encoded protein, by appropriately changing the nucleotide sequence.

In sum, mutability can apparently be altered either locus by locus or for the genome as a whole. How strong is selection for mutability? Consider, first, alleles of different stability at the same locus. Suppose our locus has four alleles, *AB, Ab, aB,* and *ab*, where *AB* mutates to *aB* and vice versa at a rate u, while the rate for *Ab* and *ab* is u'. Suppose also that selection affects *A* and *a* as in earlier paragraphs. If the environment changes every n generations, selection eliminates a proportion $u - (1/n)$ $\log(u/s)$ of the B-bearers, and $u' - (1/n)\log(u'/s)$ of the b-bearers: the selective advantage of B over b is therefore

$$u' - u + \frac{1}{n}(\log\frac{u}{s} - \log\frac{u'}{s}) = u' - u + \frac{1}{n}\log\frac{u}{u'}.$$

If the environment changes every hundred thousand generations, the optimum mutation rate is 1/100,000, but an allele whose mutability is three times too high or seven times too low has a selective disadvantage of only 1/100,000: selection to adjust the mutability of an individual locus is very weak.

What if A and B occupy loci on different chromosomes? Selection affects B only through the mutants it has caused, and to which it is still linked. Each generation a proportion u of the B-bearers cause mutants. Recombination requires two generations, on the average, to separate a mutant from the gene which "created" it. At equilibrium, when most mutants are harmful, selection eliminates a proportion s of the mutants each generation, so selection against the mutants they have caused eliminates a proportion $2us$ of the B-genes each generation. When the environment changes, B hardly benefits, since it is so rapidly dissociated from its mutants. Thus the average selective advantage of B over b is $2(u' - u)s$: an allele can never derive advantage from increasing the mutability of other loci. In sexual organisms, selection eliminates "general mutators" increasing the mutability of the entire genome, even when they benefit the species: although selection through the mutants at any one locus is slight, the aggregate can be rather strong.

A population's genetic system should function as a filter, distinguishing aspects of the environment which change too frequently to be worth following from longer-term trends: in short, a population should remember something of the past in responding to the present. If changes occur so rarely that the population does forget one before the next occurs, it may be extinguished as a result. The long-lived redwood suffered grievously from the recent glaciations because it had lost its dispersal powers and could not reclaim its former domains when the glaciers retreated, whereas firs and spruces recovered without difficulty. Many a population has become asexual to preserve a genotype suited to prevailing conditions, sacrificing the adaptability required to survive future changes.

Populations short-lived enough to be affected by the fluctuations from season to season or from year to year are likely to survive broader changes than populations to which the environment appears stabler. Often a group of animals evolves large size, apparently specializing to stable conditions, only to become extinct. The great dinosaurs died, while lizards and shrews lived on; mammoths and ground sloths failed to survive a human explosion which barely daunted mice. The history of the Cenozoic is littered with mammalian groups which evolved great size and then died out: almost always the supply of large mammals is renewed from smaller stock, never the reverse. Why should this be? A large animal is more sensitive to change because it depends on more aspects of its environment. An Indian rhinoceros requires a pond, a mud wallow, and several sorts of vegetation in its territory, any one of which can support a mouse or a turtle. An equilibrium species is more sensitive to change because it has less capacity to multiply, and cannot withstand as great a selective mortality as an opportunistic species.* If conditions begin to change in an unheard-of way,

* The total number of individuals eliminated to substitute one allele for another, say a for A, is the sum of the number eliminated each generation. Selection against a-bearers eliminates a proportion sq of each generation, where q is the a's frequency: if that generation contains N individuals, selection against a eliminates Nsq of them. The total number eliminated by one substitution is

equilibrium species will die while opportunistic species, finding the world emptied of competitors, can easily support a selective mortality which merely replaces the mortality consequent to the search for new opportunities. Mice and 'possums have litters of a dozen, while a horse has only one foal at a time; small wonder the larger animal is more sensitive to change.

The events populations cannot remember are probably the most important for us: evolution would be much less interesting could populations remember all their past. The German paleontologist Schindewolf commented that we rarely see a species change during the course of its fossil record. A species seems to appear, persist for a while, and then disappear, as if it adapted quickly to its niche and became stuck in an evolutionary rut. Rarely does a species expand its niche by direct competition: selection tends to minimize competition among different species. Extinctions provide most of the opportunities for evolution, and permit comparison of different basic designs. The ichthyosaur died out long before the porpoise evolved, and the extinction of the dinosaurs was a necessary prelude to the evolution of modern mammals. Without such crises, evolution would soon stop; without change, there would be no progress.

Problem

During the last 250,000,000 years, pelecypods have slowly been replacing brachiopods: the number of species of each at various times are tabulated below:

$$\int Nsq(t)\,dt \cong -N \log \frac{u}{s},$$

where N is population size and u/s is A's frequency when the substitution began. A population whose environment initiates k gene substitutions each generation suffers $kN \log (u/s)$ selective deaths per generation as a result. A substitution every generation kills seven times as many individuals as survive to reproduce: a population would have to be able to multiply eightfold every generation to withstand such mortality.

Period	*Permian*	*Triassic*	*Cretaceous*	*Tertiary*	*Present*
Age of Period (Years)	250,000,000	200,000,000	70,000,000	55,000,000	0
Brachiopods	7,000	3,000	2,000	1,000	250
Pelecypods	3,000	6,000	8,000	9,000	10,000

a) Suppose that there are always 10^{12} clams and brachiopods in the world, and that each individual lives a year, and suppose that the relative numbers of clams and brachiopods are reflected by the numbers of species of each. Considering clam and brachiopod as two alleles in a population, what is the average selective advantage of clams over brachiopods, per generation

1) from the Permian to the present?
2) from the Triassic to the Cretaceous?
3) from the Tertiary to the present?

b) Suppose now that selection acts between species rather than individuals, and that when a species (either clam or brachiopod) goes extinct its niche is more than likely to be filled by a clam. Clam and brachiopod species live ten million years, on the average. What is the selective advantage of clams over brachiopods, per ten million years

1) from the Permian to the present?
2) from the Triassic to the Cretaceous?
3) from the Tertiary to the present?

Solutions:

a) The selective advantage s required to alter the ratio of clams to brachiopods from 3/7 to 40 in 250,000,000 years is given by the equation

$$(3/7)(1+s)^{250,000,000} = 40,$$

$$s = 1.8 \times 10^{-8}.$$

Similarly, the selective advantages for 2) and 3) are 5.3×10^{-9} and 2.7×10^{-8}, respectively.

b) The selective pressure per ten million years is ten million times greater than the selective advantage per year, or .18 for 1), .053 for 2), and .27 for 3).

Bibliographical Notes

This chapter attempts to interpret Levins's ideas on the ways genetic systems should cope with change. The briefest summary of these ideas is his paper, "Mendelian Species as Adaptive Systems," pp. 33-38 of *General Systems*, vol. 6, 1961. He expands them in a series of papers, "Theory of Fitness in a Heterogeneous Environment": see especially Part I, "The Fitness Set and Adaptive Function," pp. 361-373 in the *American Naturalist*, vol. 96, 1962, which defines a formalism for discussing adaptation and discusses the conditions favoring clines,* polymorphisms, and other sorts of genetic variation; Part II, "Developmental Flexibility and Niche Selection," pp. 75-90 of the *American Naturalist*, vol. 97, 1963, which includes a stimulating program for testing his ideas; Part III, "The Response to Selection," pp. 224-240 of the *Journal of Theoretical Biology*, vol. 7, 1964, an exceptionally difficult paper showing when a population should respond to environmental change; and Part IV, "The Adaptive Significance of Gene Flow," which discusses the proper adjustment of migration rates. Levins amplifies his ideas further in his book, *Evolution in Changing Environments*, Princeton University Press, 1968. All these works are difficult, largely because of the originality of Levins's approach. The book is sometimes clearer than the papers, but is less balanced in tone: just as

* Any characteristic of a population (such as color or average size) which changes evenly from one part of the population's range to another is said to exhibit a *cline*.

D'Arcy Thompson sometimes supplies artificial and unconvincing mechanical explanations, so Levins sometimes strains adaptive explanations. C. L. Remington argues with immense learning that most hybridization is a result of historical accident (and presumably is not yet precisely tailored to the needs of the species involved): see his paper, "Suture-zones of Hybrid Interaction between Recently Joined Biotas," pp. 321-428 of *Evolutionary Biology*, vol. 2, 1968.

Levins's work sets in proper perspective an old controversy between geneticists believing that genetic variability is stored primarily as recurrent mutants and those emphasizing the prevalence of balanced polymorphisms. See, for example, H. J. Muller, "Our Load of Mutations," pp. 111-176 of the *American Journal of Human Genetics*, vol. 2, 1950, and chapter 9 of E. Mayr, *Animal Species and Evolution*, Harvard University Press, 1963. The controversy acquires emotional overtones when believers in "mutational" variation drift toward belief in an ideal "wild-type" genotype, which should perhaps be imparted to everyone through the good offices of eugenics bureaus, while advocates of balanced polymorphism emphasize the intrinsic virtues of diversity.

The ecology of micro-organisms is discussed by G. E. Hutchinson in "The Paradox of the Plankton," pp. 137-145 of the *American Naturalist*, vol. 95, 1961.

M. Kimura calculates optimum mutation rate in his paper, "On the Evolutionary Adjustment of Spontaneous Mutation Rate," pp. 23-34 of *Genetical Research*, vol. 9, 1967.

A whole symposium has been devoted to the genetics of colonizing species: see Baker and Stebbins, eds., *The Genetics of Colonizing Species*, Academic Press, 1965.

The work Cain, Sheppard, and others have done on *Cepaea* is summarized in E. B. Ford, *Ecological Genetics*, Methuen, 1964; Dobzhansky discusses the chromosomal polymorphisms of *Drosophila* in *Genetics and the Origin of Species*, Columbia University Press, 1951.

Crow discusses the advantages of an additive genetic system in his paper, "Genetics of DDT Resistance in Drosophila," pp. 408-409 of the *Proceedings of the International Genetics Symposia*, 1956 (supplemental volume of *Cytologia*, published in 1957).

An experiment lowering bacterial mutability is reported in Zamenhof, Heldenmuth and Zamenhof, "Studies on Mechanisms for the Maintenance of Constant Mutability: Mutability and the Resistance to Mutagens," pp. 50-58 of the *Proceedings of the National Academy of Sciences*, vol. 55, 1966. The ways bacteria correct replication errors are discussed in Hanawalt and Haynes, "The Repair of DNA," pp. 36-43 of *Scientific American*, vol. 216, no. 2, 1967.

Benzer discusses "hot spots" in his paper, "The Elementary Units of Heredity," pp. 70-93 of W. D. McElroy and B. Glass, *The Chemical Basis of Heredity*, The Johns Hopkins University Press, 1957.

Selection for and against mutators is discussed in E. Leigh, "Natural Selection and Mutability," pp. 301-305 in the *American Naturalist*, vol. 104, 1970.

Dillon Ripley discusses the requirements of Indian rhinos in his paper, "Territorial and Sexual Behavior in the Great Indian Rhinoceros: a Speculation," pp. 570-573 of *Ecology*, vol. 33, 1952.

Haldane explores the notion that populations can withstand only a limited selective mortality in his paper, "The Cost of Natural Selection," pp. 511-524 of the *Journal of Genetics*, vol. 55, 1957. This notion does not imply a limit to the rate of evolution (despite Haldane's views to the contrary): a new mutant spreads only if it is more reproductive than its alleles, in which case the increased reproductivity it confers enables the population to "pay for its substitution." (There is no evidence in the fossil record for a genetic limit to the rate of "progressive" evolution: in spite of their long generations, elephants have evolved as fast as any group.)

A related idea is the principle of "minimum genetic load," which is the principle of maximum fitness stated backwards. The genetic load of a population is defined as the difference between the fitness of its optimum genotype and the fitness of the population as a whole. The definition implies a single optimum genotype: this is useful in speaking of recurrent harmful mutations (see H. J. Muller, "Our Load of Mutations," pp. 111-176 of the *American Journal of Human Genetics*, vol. 2, 1950), but those who use it to limit the prevalence of balanced polymorphisms, measuring a population's fitness against the imaginary standard of the ideally heterozygous individual, are led to doctrines to which, in view of the available evidence, it becomes ever harder to subscribe.

Index of Authors and Subjects